Manufacturing cells: control, programming and integration

Manufacturing cells: control, programming and integration

Edited by

*David J. Williams, Department of Manufacturing
Engineering,
Loughborough University of Technology,
Loughborough, UK*

and

*Paul Rogers, Department of Mechanical Engineering,
The University of Calgary,
Calgary, Alberta, Canada*

Butterworth-Heinemann Ltd
Linacre House, Jordan Hill, Oxford OX2 8DP

 PART OF REED INTERNATIONAL BOOKS

OXFORD   LONDON   BOSTON
MUNICH   NEW DELHI   SINGAPORE   SYDNEY
TOKYO   TORONTO   WELLINGTON

First published 1991

British Library Cataloguing in Publication Data
Manufacturing cells: control, programming and
integration.
  I. Williams, David J.   II. Rogers, Paul
  670.285

ISBN 0 7506 0235 X

Library of Congress Cataloguing in Publication Data
Manufacturing cells: control, programming, and integration / edited
  by David J. Williams and Paul Rogers.
    p.   cm.
  Includes bibliographical references and index.
    ISBN 0 7506 0235 X
    1. Production control–Automation.   2. Computer integrated
manufacturing systems.   I. Williams, David J.   II. Rogers, Paul.
TS155.8.M33   1991
670.42′75–dc20   91–24750
                                                                 CIP

Printed and bound in Great Britain

Contents

Preface and acknowledgements

This book is intended to examine the issues surrounding the integration of manufacturing cells and small manufacturing systems. It arises from a research interest of ours that began in the mid-1980s. When considering what work to do ourselves it became apparent that there were a number of groups that had been working in this area for some time. These groups have also continued to work in the area since that time.

We have attempted to draw together the work of these groups and present it, with a brief introduction, in a single volume. Readers active in the area will recognize that some of the work has already been published in the journal literature. We have, however, sought to persuade the authors of each piece not only to put in material that puts forward their latest thinking in the area, but also to discuss the 'history' of their work to show why they took their present technical position. We hope that this approach will be valuable to old and new workers in the area.

Preparing any book is a long job. Preparing an edited text is less effort for the editors than it is for the authors of each contribution! Our thanks must go to each of the contributors for their part of the book, for the preparation of the text, for responding so positively to our 'editorial' comments and for being so patient with us while waiting for the text to be put together.

David Williams, Loughborough University, 1990
Paul Rogers, Purdue University, 1990

Contributors

D. A. Bourne, PhD
Robotics Institute, Carnegie Mellon University, Pittsburgh,
Pennsylvania, USA

D. Burnage
Reflex Manufacturing Systems Ltd, Crawley, Sussex, UK

T.-C. Chang, PhD
School of Industrial Engineering, Purdue University, West Lafayette,
Indiana, USA

I. A. Coutts, BSc
Department of Manufacturing Engineering, Loughborough University
of Technology, Loughborough, Leicestershire, UK

N. A. Duffie, PhD
Manufacturing Systems Engineering, University of Wisconsin-Madison,
Madison, Wisconsin, USA

J. D. Gascoigne, BSc
Department of Manufacturing Engineering, Loughborough University
of Technology, Loughborough, Leicestershire, UK

A. Hodgson, BTech, MSc
Department of Manufacturing Engineering, Loughborough University
of Technology, Loughborough, Leicestershire, UK

A. Jones, PhD
National Institute of Standards and Technology, Gaithersburg,
Maryland, USA

T. Jones
Reflex Manufacturing Systems Ltd, Crawley, Sussex, UK

I. S. Murgatroyd, BSc
Department of Manufacturing Engineering, Loughborough University
of Technology, Loughborough, Leicestershire, UK

P. Rogers, MA, PhD
Department of Mechanical Engineering, The University of Calgary,
Calgary, Alberta, Canada

A. Saleh, PhD
AT & T Bell Laboratories, Holmdel, New Jersey, USA

R. H. Weston, BSc, PhD, DSc
Department of Manufacturing Engineering, Loughborough University of Technology, Loughborough, Leicestershire, UK

D. J. Williams, BSc, PhD
Department of Manufacturing Engineering, Loughborough University of Technology, Loughborough, Leicestershire, UK

1 *An introduction to the manufacturing cell and its integration and control*

David J. Williams

1 THE MANUFACTURING CELL

Automated manufacturing cells are the practical building blocks of CAM and CIM systems, and are important in their own right as islands of automation. Such cells usually consist of a number of closely cooperating different machines coordinated and controlled by a supervisory computer.

Cells have been constructed for a wide range of applications: prismatic ('box-like') metal cutting using machining centres, revolute (cylindrical) metal cutting using turning centres, electronic assembly using industrial robots and placement machines, and for a wide range of robotic processing applications. Increasingly cells are being introduced into the manufacturing plant and progressively integrated into larger plant-wide automation and factory control systems.

The integration task has been identified as one of the major bottlenecks in the construction of effective manufacturing systems. The computational complexity of this systems engineering task has given many problems to manufacturing organizations most of which have more conventional mechanical engineering or industrial engineering skills.

It is the purpose of this book to focus on techniques that allow the effective and fast integration of manufacturing devices into automated manufacturing cells and then, in turn, allow the integration of such cells into the complete manufacturing system. These techniques must be understood to allow the manufacturing community worldwide to exploit the full potential of the manufacturing cell both as a standalone tool and as part of a larger system.

2 WHAT DOES THIS BOOK ADDRESS?

The mechanical design of cellular systems is becoming well understood as is the selection of the components that should be manufactured within a cell. It is, however, necessary to turn this mechanical design into a functioning system that allows the production of the components. The problems that must then be resolved are the integration, coordination and control of the cell devices.

The key to the resolution of these problems is the representation and transportation of information between the system elements. Many chapters will return again and again to the themes of integration architectures, the use of data and databases and the response of the cellular system to unforeseen events – the recovery problem.

The interest of the editors in this field began while they were working in the Manufacturing Engineering Group of the University Engineering Department, Cambridge. During their work together it became apparent that there were a number of laboratories, worldwide, that had made significant contributions. This book aims to draw together contributions from these laboratories, each contribution representing not only a view of a key technology, but a view of that technology that has been refined by continuous experiment and peer-reviewed publication. Each chapter therefore represents an overview of work in each laboratory written with the experience of hindsight and an understanding of the longer-term research opportunities that arise.

This book is therefore intended to present a broad view of the present thinking on the resolution of the cell and system level integration problem. Each chapter will also indicate unanswered questions to spark the interest of the researcher. The later chapters present work that is perhaps more speculative in nature and that may provide the basis of factories of the less immediate future.

3   WHAT IS IN THE BOOK?

The purpose of this chapter is to review briefly the concepts of cellular manufacture and introduce each of the chapters, alerting the readers to that which should be drawn from each chapter. To begin this task the order of the material within the book will be discussed.

The body of the book begins with a report of the work of the Systems Integration Group of Loughborough University, this chapter focusing on tools that ease the integration of information between shop-floor devices and tasks. The next chapter, from workers at the Advanced Manufacturing Research Facility at the National Institute of Standards and Technology, describes the evolution of their systems integration work. Jones and Saleh also describe present thinking on the implementation of reactive shop-floor control systems at NIST. David Bourne then describes the cell management language, CML. This chapter concentrates on the practical implementation of lower-level software systems that provide effective communication between manufacturing machines and allow them to be coordinated to perform a useful function.

These three chapters summarize a major body of research work worldwide on the integration of large and small systems. The chapter that follows, from Burnage and Jones from Reflex, shows how these research efforts and tools are being echoed by commercial products.

The next two chapters look further into the future and focus on other complementary issues. Chang from Purdue University addresses process planning both for the cutting machine and the inspection machine in the cell. Chang also describes the coordination of the plans generated for these machines with the actions of other machines and people within the cell. The chapter by Duffie demonstrates the coordination of a larger group of machines by negotiation rather than the more prescriptive hierarchical methods encountered to date.

The closing chapter by the co-editor, Paul Rogers, draws together the underpinning control formalisms within each of the chapters into a more structured framework highlighting the use of object-oriented techniques, and projects some of the future research issues in the area.

4   CELLULAR APPROACHES

It is now appropriate to review briefly the functionality of the cell within manufacturing. There are many ways of considering the manufacturing cell.

The cell can be viewed in terms of the items that it usually contains; from this perspective we can see it as a small number of closely cooperating machines. Close cooperation can include, for example, the sharing of dimensional data between a measuring machine and a machine tool, the sharing of a workspace between a robot and a turning machine and the cooperation inherent in the use of a team of people to run the cell. Such machines often work in parallel, some of the machines in the cell carrying out manufacturing tasks at the same time. The cell can also be viewed using a biological analogy as the smallest autonomous unit capable of sustained production.

Many of the present generation of definitions take a factory automation (FA) or automated manufacture view of cellular manufacture. It must be recalled that the cellular concept arose from the application of group technology. Group technology (GT) was an important precursor to the design of automated manufacturing cells because it grouped different sorts of conventional machines together into 'GT cells'. Each of these cells was designed to produce a 'group' or parts family. A parts family is defined as a set of parts that require similar machinery, tooling, machine operations and jigs and fixtures. The parts themselves are handled between machines manually and the machines themselves are controlled by an operator.

The parts to be made within a GT cell are selected to be similar either by coding or by visual inspection. Group technology cells can reduce work-in-progress and generally increase the operating efficiency of small batch manufacture by reducing handling and transport costs. The design discipline implied by the grouping activity also reduces the proliferation of very similar but different product designs which fulfil essentially the same function. GT cells are now, once again, widely applied in just-in-time manufacturing environments, as they give those running the cell ownership of the complete production of a group of parts.

The range of parts that an automated manufacturing cell produces is, however, usually less than that in many conventional GT cells and also there are frequently less machines in the cell. This has often arisen from a necessity to minimize the mechanical and control complexity of the automated installation. In factory automation, the cell has usually been built to get the optical performance from a very expensive machine tool and the balance of the machinery serves this machine so that it has minimum idle time.

The automated cell requires autonomy. Autonomy, in this case, means minimally manned manufacture for a sustained period – in the foreseeable future the completely unmanned factory is not practical nor desirable nor financially justifiable. A human operator will always be required for system patrol, maintenance and system recovery in some circumstances – the degree of manning must be consistent with economic and reliable system operation.

This autonomy requires automated or programmable processes; automated handling to and from these processes; automated quality control or inspection of the performance of the processes together with rapid feedback of any necessary changes to each process; supervisory control and sensors to monitor and detect the cell condition and to decide the next activity. Many industrial cells do not have the integral inspection facility, this being carried out as is usual in more conventional facilities by a patrolling operator. It is usual in factory automation, however, to take good account of the capability of the manufacturing process and to design cells with a 'right-first-time' approach.

5   ADVANTAGES AND DISADVANTAGES OF CELLULAR APPROACHES

The primary advantage of the cell when building systems for factory automation is that it reduces system mechanical complexity. This reduction allows easier financial justification, less technological risk, easier installation and commissioning, and allows standalone islands of automation to be built if these are justified by business considerations.

The advantages of the manufacturing cell are: reduction of the necessary 'organizational' control involved in bringing materials to a cell; reduced handling of parts; reduced set-up time, often by purpose design of jigs and fixtures; reduction of work-in-progress and inventory by careful attention to buffer design; and a reduction in the need to expedite (progress-chase) between machines.

The most fundamental disadvantage of the cell is that of the island of automation. Even the most efficient automated cell, if it is isolated, may just move the manufacturing problem somewhere else in the manufacturing organization. A significant, and frequently encountered, example of this is the build-up of work-in-progress before or after the cell, which does not increase overall system efficiency. An isolated cell must therefore be part of a well-considered overall manufacturing facility design.

6   CONTRIBUTIONS

It is the focus of this book to address the software issues that arise in manufacturing and the construction and operation of automated manufacturing cells, so the contribution of each chapter will now be discussed in turn.

6.1   The Contribution of Weston and his Colleagues

The book essentially begins with the chapter written by the Systems Integration Group led by Professor Richard Weston in the Department of Manufacturing Engineering at Loughborough University of Technology. By focusing on the technologies needed to achieve flexible integration and generate economically reconfigurable cell controllers that fit within larger CIM systems, the authors develop a requirements specification for such systems.

They then turn to present their own solution, the AUTOMAIL system, which is targeted to reduce the systems engineering effort inherent in the integration process. AUTOMAIL allows modular manufacturing entities – whatever is required to be manufactured, be it machines or people – to be integrated. The software system enables the construction of cell and system controllers that can transfer information to and within wider systems, information integration being identified as the major function of any integrating architecture. The solution framework generated by Weston and his group also permits efficient controller reconfiguration to accommodate the changes in requirements inherent in the evolving manufacturing world.

Within this chapter the Loughborough Systems Integration Group also review the concept of the three-schema architecture. This approach identities the three different views that can be taken of any integration architecture: the information architecture which structures the data within the system, the network architecture which describes the topology of the interconnection between system elements; and the application architecture which structures the lower-level software application implementations that manipulate the data structures.

In contrast to the work of many of the other authors in the book such as Bourne and Jones and Saleh, who use a purely mechanical manufacturing test bed, the demonstration application for the system is electronic assembly. All of the systems discussed in the book provide generic solution frameworks, but the authors have demonstrated the functionality of their particular systems in practical situations to verify their performance.

6.2   The Contribution of Jones and Saleh

Chapter 3, written by Dr Albert Jones of the Advanced Manufacturing Research Facility of the National Institute of Standards and Technology with Dr Abdol Saleh of AT & T Bell Laboratories, describes some of the current thinking on manufacturing control at NIST. This thinking is the result of many years' work on the construction of hierarchical control systems for manufacturing and robotics with the aim of producing manufacturing control systems that are adaptive.

The chapter focuses particularly on the representation of the functionality required to allow the distribution of integration and control activities and the design of cell controllers within such a system.

Hierarchical models of manufacturing such as those pioneered by the AMRF identify the cell (or workstation in the AMRF definitions) as the next level of integration above single machines. From this viewpoint a cell characteristically consists of a number of machines of different types supplied by different manufacturers, an approach explicit in the work of Bourne that follows.

6.3   The Contribution of Bourne

This chapter, written by Dr David Bourne of the Robotics Institute, Carnegie Mellon University, presents the development of the cell management language (CML), an environment that enables different devices to be integrated into cells and allows the programming of their interaction in a high-level language. The major contribution of Bourne's work was the recognition that it was necessary to provide an environment of grammatical language tools to assist the interfacing and language translation needed to create a working system. The aim of the work was to provide tools that were easily usable by all involved in manufacturing, not solely by those in supporting computing services. These tools do not therefore require traditional programming skills.

This work has been fundamental to the thinking of many researchers and demonstrates the power of the application of tools developed by computer scientists to manufacturing problems. CML, used in factories since 1984, is one of the first powerful real-time rule-based tools used in manufacturing and one of the first systems to attempt to allow the system to accommodate abnormal conditions.

6.4   The Contribution of Burnage and Jones

The work described so far has been carried out in universities or in government-funded laboratories, but the work contained in Chapter 5 has been carried out in a commercial company, Reflex Ltd (now a company of Rolls-Royce plc) with the support of the UK Department of Trade and Industry.

This chapter describes the product CIMPICS, a programming environment that allows the user to synthesize CIM systems using graphical tools to capture the logic of device interactions and to create the interfaces to machines generating these interactions. The Reflex system is based upon the graphical programming language Grafcet, which allows easy expression of the logical interaction of concurrent processes. Grafcet is then extended by a complementary Reflex-designed graphical programming system, Graflex, which allows the generation of procedures to control each of the actions implicit in the logical system designed in Grafcet. Each Graflex application is synthesized from a library of object-like modules. CIMPICS also contains graphical tools specific to the integration tasks such as area and cell scheduler configuration tools, interfacing and interface protocol specification tools. Special computational facilities to allow the cell and area control tasks to proceed effectively include a real-time database and an interprocess communications system.

The description of the integration tool is followed by an extended description of a CIMPICS installation at Rolls-Royce Sunderland, which is part of a larger plant-wide manufacturing control system. The chapter describes the technical detail of the installation from machine integration, through cell and area scheduling, to links to CIMPICS external systems like MRPII. It also discusses the business reasons for building the computer system and the approaches taken in its design and project management.

6.5   The Contribution of Chang

In most metal-cutting applications a key part of the cell is the numerically controlled metal-cutting machine tool itself. One of the major manufacturing bottlenecks is the necessity to generate the control code for such machines. The chapter by Dr T. C. Chang describes the quick turnaround cell (QTC), a major project within the Intelligent Manufacturing Systems Engineering Research Center at Purdue University. The QTC tackles the problem of process planning and the generation of manufacturing instructions for cutting, cell control, fixturing and inspection from an appropriate geometric description of the part, at the machine.

The emphasis of the work is to reduce the design-to-product lead time and increase system flexibility, the fast generation of process plans and part programs both for stock preparation and final machining being fundamental to both. The cell constructed at Purdue is primarily intended as a research test bed that emulates the very small batch production required for machine maintenance functions. The cell also uses the human, where he or she fits best, within the design function, providing the part designer with a user-friendly environment within which to carry out his or her work.

The planning approach used by Chang utilizes a combination of feature extraction and manufacturing-based feature modelling to generate and reason about design descriptions. This feature information is then passed to the rule-based process planning system which generates cutter path files. In the QTC an operator is used to load and unload parts; this operator is advised on the tools and parts loaded into the equipment by the cell control system. Final inspection is carried out using a vision system guided by the process planning information to check for visible part features.

6.6   The Contribution of Duffie

The cells described above have assumed that there will be a strictly hierarchical master–slave structure used for the integration and coordination of the manufacturing devices. Dr Neil Duffie and his colleagues at the University of Wisconsin–Madison have recognized that there are other ways of coordinating distributed systems and have implemented a novel negotiation- or cooperation-based control system to control a cell within their laboratory.

The manufacturing integration task is the integration of a large heterogeneous collection of manufacturing entities into a system and with the outside world. Duffie and many others consider that the more novel negotiation-based architectures are likely to be much more fault tolerant than other architectures. In addition they feel that such architectures will allow reduced software costs because of their reduced complexity and improved modularity.

The chapter describes the cooperative philosophy worked upon since the early 1980s, together with the design and integration of a number of generations of Wisconsin–Madison modular drive systems and manufacturing cells, and compares the effort needed for integration of the most recent using a number of coordination paradigms. Constructed cells have included a part-oriented heterarchical control system and a larger multi-cell system for automated plastic mould production, some of the cells in this larger system being emulated by simulations.

6.7   The Contribution of Rogers

The closing chapter by Dr Paul Rogers restates the major issues in manufacturing control and overviews the possible control and integration strategies in a more formal framework. The chapter then focuses on the use of object-oriented modelling and its relationship to other techniques. Rogers closes by projecting future research directions, particularly the extension of object-oriented approaches to truly distributed cell control and simulation.

One of the particular aims of the chapter is to discuss one of the most promising technologies to allow the production and reuse of modular software structures – object orientation. The message-centred style of object orientation makes it attractive for the construction of the coordination software required for cell control.

7   CLOSING COMMENTS

Readers will become aware of a number of underlying themes as they progress through the book. Key amongst these are the representation and distribution of information within the cell, the hardware and software architecture within the cell and its interface to the external system, and the necessity to ensure that cells are flexible and able to adapt to changes in their environment.

Readers will also recognize that the manufacturing community is increasingly learning from the artificial intelligence and software engineering communities. This has resulted in the new generation of manufacturing engineers we can see working on the projects presented within this book. These engineers are becoming as literate in computer science and its related disciplines as they are in manufacturing practice.

2 *Generally applicable cell controllers and examples of their use*

R. H. Weston, I. Coutts, A. Hodgson, S. Murgatroyd and J. D. Gascoigne

It is not the intention of this chapter to present alternative scheduling techniques, sequencing algorithms or other tactical issues. The issues of primary interest relate to the problem of achieving flexible integration. Methodologies are presented which enable the implementation of flexible, economically reconfigurable cell controllers which can cope with the dynamic environments of modern manufacturing organizations.

In this chapter, we view the cell controller as being only one element of a computer-integrated manufacturing (CIM) system. By adopting this stance, the required features of future cell controllers are identified so that they can be included in a highly flexible and cost-effective manner into integration schemes with any target manufacturing domain in mind.

To develop and illustrate the concepts involved, solutions to integration problems are advanced and related to electronic product manufacture. Various cell control implementation schemes are described in order to provide a detailed understanding of necessary integration mechanisms and hence a requirements specification for a set of integration tools which can be widely applied.

1 FACTORY INTEGRATION AND THE ROLE OF CELL CONTROLLERS

In order to remain competitive, companies operating in the various industrial domains need to adopt market-driven manufacturing methods and new technological tools. However, an array of market needs and environmental constraints exist which result in an almost infinite number of potential solutions. As product life cycles reduce, manufacturing organizations need to become more responsive and can increasingly be considered to represent dynamic systems subject to frequently changing requirements.

By utilizing information from computers used in product design, manufacturing organization and control and shop-floor production, a cell controller can be considered to be an instrument for achieving a degree of factory integration leading to CIM. It provides operators and management with supporting information and decision-making capabilities, which in turn can lead to increased productivity, resource utilization and product

Table 2.1 *Typical interacting entities*

Higher-level MEs:
  Production planning & control system
  Materials tracking system
  Quality system
  Information management & storage system
  Costing & financial systems
  Scheduling entities
  Human interface elements
  Shop controllers

MEs at the cell level:
  Other cell controllers
  Materials control systems
  Production monitoring entities
  Diagnostic/condition-monitoring entities
  Human interface elements

Lower-level MEs:
  Robots
  Programmable logic-controlled machines
  Digitally controlled process loops
  CNC machines
  Assembly machines
  Inspection systems (including machine vision)
  Workpiece transport systems
  Tool transport systems
  Operator interfaces

quality. As such it should be a significant factor in achieving increased profitability.

Although it is clearly necessary to reference technical advances made and available systems when determining the required features of future cell controllers, it is also essential that this class of manufacturing system component is not considered in isolation. The descriptions of cell control schemes in chapters of this book serve to illustrate the complexities involved in producing systems which can be applied in any manufacturing scenario where widely varying integration requirements exist. Any cell controller which purports to being generally applicable must demonstrate sufficient functionality to allow it to interact with a vast array of manufacturing entities (MEs), the term ME being used for convenience to denote 'whatever is required to accomplish one or more manufacturing activities'. Table 2.1 lists typical MEs with which a cell controller may interact. This list is not intended to be exhaustive, and serves only to illustrate the nature of MEs, which can comprise some combination of computer hardware, software and interface electronics as well as plant actuators and sensors. Furthermore in

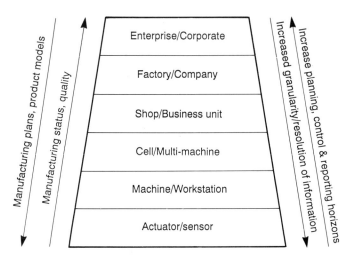

Figure 2.1 *Factory hierarchy. This organizational structure is favoured by the authors but it is fully recognized that (1) not all levels will exist in a given organization, (2) certain MEs will straddle organizational boundaries, and (3) many organizational models exist which reflect the stance taken when attempting classification*

this context, people (be they directors, engineers, managers, supervisors, or shop-floor workers) can be considered to be MEs, this point being made to emphasize that the use of mechanized automation is not a prerequisite of CIM systems. In this chapter, all MEs will be considered to be supported in some way by a computer system (e.g. an assembly operative by a computer terminal or a manager/supervisor by decision and information support tools). In this way electronic information can facilitate high-speed interaction amongst MEs.

Although there are wide functional variations between MEs, they all 'process' information so that generally their actions will result in software-based state changes (e.g. reformatted information). Certain MEs (such as a robot, or an operator/decision support terminal combination) also function to produce real-world state changes. Thus an interpretation of the role of a cell controller is that it should manage the interactions between MEs under its charge, thereby changing states (software or real world) according to some plan of action which may be generated by an ME higher up the manufacturing hierarchy (this hierarchy may involve the functional levels depicted by Figure 2.1) (Van Dyke Paranuk and White, 1989; Ottawa, 1986).

A cell controller and its attendant lower-level MEs can be considered to be a single composite ME when viewed externally by a higher-level entity such as a shop controller. The cell controller itself can also be viewed internally as comprising a number of more primitive MEs. The need for these different views of MEs arises when considering organization, integration and implementation issues, and suggests a mechanism whereby generalization of the factory integration problem can be achieved.

In this chapter we concentrate primarily on cell control MEs by considering specific examples of their application and drawing out general observations. However, the concepts described can equally well be applied at any level in the factory hierarchy.

2 IMPORTANT CELL CONTROLLER DESIGN CONSIDERATIONS

When integrating the activities of a set of MEs to perform a manufacturing task, a range of system design issues needs to be considered. This section introduces briefly some of the more important issues involved. A basic understanding of the relevance of these issues is required before moving on to the descriptions of actual cell control implementations.

2.1 Operational Flexibility

When considering a specific cell control application, an appropriate level of operational flexibility will be required, dependent on the range and type of products to be made and anticipated changes in operation conditions.

To increase this operational flexibility, significant advantages can be gained by achieving software control of both 'real-world' and 'software-based' states. Examples of the former are programmable tool changing at a robot arm, software-based adjustment of vision system lighting conditions and camera position, and programmable selection of pallet sizes. At the machine level, cost and processing time constraints may limit attainable increases in operational flexibility through accomplishing programmable changes of real-world states. Significant advantages can also be gained by generating software-based state changes to direct operators when changing corresponding real-world states. The loading, selection and instantiation of robot task programs, vision system image processing algorithms and conveyor routing sequences are also typical of software-based state-change requirements where subsequent processing of those states (e.g. in a robot controller) results in task execution. Whichever class of state change, the primary objective is one of reducing system changeover times (thereby facilitating small batch production with high levels of resource utilization) and improving the system robustness (thereby achieving corresponding improvements in product quality).

However, in realizing sufficiently high levels of operational flexibility at the machine level, it is also necessary to employ cell controllers which can maintain the required flexibility when defining and controlling the interactions between MEs under their charge. Such controllers must also maintain appropriate operational flexibility where they are required to interact with other high-level MEs.

2.2   Configurability

Using tailored hardware and software solutions, a group of MEs can be combined to perform a single, specific manufacturing application. However, in this instance, redefinition of the application (which may include removing one or more MEs, adding one or more MEs, resynchronizing the interactions of the MEs, loading new applications software on each or some MEs, etc.) would involve significant engineering effort and cost and would result in yet another specific solution. The cell controller must therefore be capable of redefining interactions of the MEs under its charge. This identifies the need for cell controllers to offer flexibility with respect to configurability, where this type of flexibility relates to the requirement to:

(a) install such controllers in a wide range of manufacturing scenarios; and

(b) modify/enhance their functionality in a given scenario as changes in markets and/or required operating practices dictate system changes.

In order to achieve this configurability in an economic manner, the following must be achieved:

(1) a consistent representation of manufacturing entities; and

(2) the availability of a set of system configuration tools.

Consider the first point: the basic functionality of an ME is dictated primarily by the tasks for which it is designed, for example a particular robot arm is designed for component insertion, a vision system for inspection, etc. Such MEs may have local computer-based processing capabilities associated with, for example, proprietary communication protocols or for acting upon sensory inputs. These capabilities will vary widely depending on the above task requirements and the approach taken by the particular ME supplier. They will not typically correspond to any general idea of a 'standard' ME, being designed by the supplier in some specific way in order to gain a competitive edge in the equipment marketplace.

It is therefore necessary to insulate the generic cell control functions from the wide variations in functionality and protocol found in typical low-level MEs. The implication of this is that these MEs are enhanced, as necessary, so that each can be treated as some form of standard 'virtual manufacturing device' (VMD). Each would possess at least a minimum level of function-command and data structures. This would allow interactions such as 'start' and 'stop' commands and the interchange of files and variables (Weston *et al.*, 1989a). The above enhancements to turn proprietary MEs into 'standard' VMDs would logically be located at the individual MEs. However, in the case of simple low-level MEs such as sensory transducers, this additional functionality might more economically be located at the cell controller as a set of processes separate from the generic cell control functions.

This approach of defining VMDs has been adopted by several standardization and research initiatives. An example of this is the Manufacturing Message Services (MMS) specification of the Manufacturing Automation Protocol/Technical Office Protocol (MAP/TOP) local area network (LAN) initiative (1988). This point is covered in more detail in the following subsection. Our view of MEs/VMDs is wider than that typically implied elsewhere by the use of the term VMD, as we also consider software entities to be legitimate MEs in their own right (Weston *et al.*, 1989b). An example of this would be a virtual or standard scheduling ME providing appropriate services to a cell controller.

The development of standard VMDs does not in itself represent a solution to the problems of developing reconfigurable cell controllers, but a prerequisite. The second of the two requirements listed earlier (i.e. a set of configuration tools) is equally important. With a consistent representation of manufacturing devices, it becomes feasible to specify and develop an appropriate set of configuration tools. These would facilitate the systems engineering tasks associated with building, testing, debugging and re-configuring the cell controller to carry out the planning, control, synchronization and information collection functions required of it.

2.3   Choice of Communication Architecture

The problems of establishing meaningful information exchange between MEs can be broken down into three separate areas as shown in Figure 2.2. The choice of physical connection between MEs must be made with due regard to availability, ease of use, environmental conditions and the hierarchical (or possibly heterarchical) nature of the cell. For example, actuator/sensor level MEs might be connected using a high-speed backplane bus (such as Multi-bus), machine-level MEs via twisted-pair cabling, and cell controllers connected via a LAN based on coaxial cabling to MEs operating at the same or at higher organizational levels in the factory.

For each physical connection a suitable protocol must be chosen. Many possible choices exist, for example RS232, RS432, Ethernet, MAP, TOP, etc. The protocol used might conform to international or *de facto* standards, or might even be custom designed by a system builder. The choice of protocol should take into account the data rates and data quantities involved and the need for time-critical data transfers.

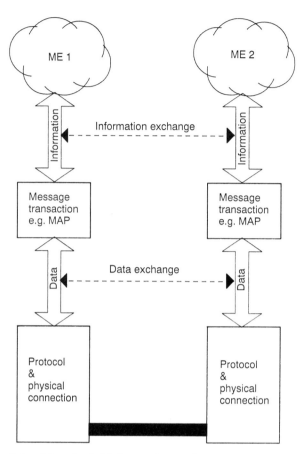

Figure 2.2 *Meaningful information exchange*

The use of 'standard' protocols is preferable, but with the current state of enabling technology, inherent processing overheads associated with a generally applicable approach may limit its use. This is particularly true of the actuator/sensor and machine levels of the manufacturing hierarchy, where short bursts of time-critical information often occur. Therefore, customized protocols will continue to be used for some time in these situations.

Having connected entities together and established a choice of communications protocol, the MEs can exchange data. However, the meaning of this data is normally ME specific. To overcome this problem, the concept of a virtual manufacturing device (VMD), introduced earlier, can be used to allow shop-floor manufacturing devices to be treated in a consistent manner, for example as a 'standard robot' or a 'standard vision system'. The MMS (Manufacturing Message Services) specification has been aimed at achieving this by providing software services for manufacturing applications which can be used over MAP LANs or, theoretically, over any communications protocol. However, MMS has yet to become available for all types of ME. Furthermore, to date, even MMS kernel services (MMS, 1987) are not widely available with respect to different computer makes and operating systems.

The current approach to the Technical Office Protocol (TOP) initiative takes a broader view of the range of entities which might be found in a manufacturing organization. Specifications are currently being advanced to provide virtual terminal, graphic, product model and electronic mail access via communication networks (Weston *et al.*, 1988).

The foregoing discussion serves to illustrate that, like a telephone system, the communication architecture provides a 'communication service' which, via a chosen set of communication primitives, can speed up the transfer of meaningful information between VMDs. Thus, highly responsive manufacturing systems can be created by allowing MEs to interact in a coordinated, timely and efficient manner. However, as is explained later, the realization of this potential is not a trivial matter and will require other 'services' so that flexible CIM systems can be treated in a cost-effective way.

3   EXPERIMENTAL CELL CONTROL SYSTEM

3.1   Cell Control Scenario

The Systems Integration (SI) Research Group at Loughborough University of Technology (LUT) has constructed a flexible assembly system which utilizes a number of typical MEs to demonstrate the principles and use of reconfigurable cell controllers (see Figure 2.3). One of the application scenarios represented by this system has been based on the assembly of various configurations of switches into a palletized tooling plate (see Figure 2.4). The set of MEs used to achieve this particular task included

(1) a Bosch pallet-conveyor system;
(2) a manually operated (but decision-supported) palletizing and loading/unloading station;
(3) an AdeptOne SCARA (Selective Compliance Assembly Robot Arm) manipulator arm;
(4) an AdeptOne area vision system (this being an integral part of a second AdeptOne robot, the manipulator not being used in this particular application); and
(5) a number of computer systems providing information and decision support capabilities for managers and supervisors.

A simple assembly task was deliberately chosen because the prime object of this work was to emphasize the actions of the cell controller rather than the

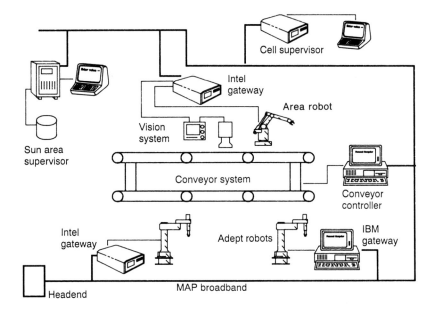

Figure 2.3   *Flexible assembly system*

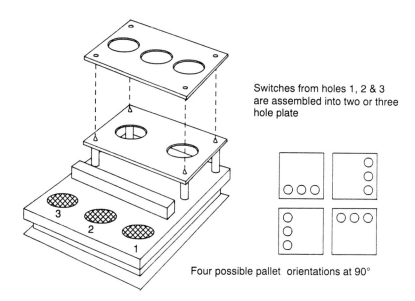

Switches from holes 1, 2 & 3
are assembled into two or three
hole plate

Four possible pallet orientations at 90°

Figure 2.4   *Switch assembly*

mechanical problems of the assembly process (tooling, grippers, manipulator actions, etc.). However, this task enabled the emulation of typical information flows around assembly cells and required the use of MEs of various functionalities to provide numerous assembly configurations.

The simple set of manufacturing actions performed by the cell are as follows. An empty pallet is driven to the loading station and the operator instructed (via a terminal) to load the pallet with a set of switches. When this task has been completed, the operator releases the pallet (by the push of a button) which is then transported to the vision inspection station. The vision system firstly determines the type of assembly (two or three switch) and the orientation of the pallet on the conveyor (one of four). The vision system then determines the position and type of each switch present. If the switches present are compatible with the type of assembly, the pallet is driven to the assembly station and assembled by the SCARA robot. If the switches are incompatible, the pallet is driven to the unloading station and the operator is instructed to unload it. Successfully completed assemblies are driven to a buffer zone of the conveyor.

3.2   The Control Architecture

In the example manufacturing scenario, the machine-level MEs were originally to be managed by a single cell controller. However, to demonstrate high levels of flexibility and to emulate integration in a wider factory environment, a decision to implement two reconfigurable cell control sites was made, with each site capable of performing the desired control function. These sites in turn would be managed by a single area controller.

In this hierarchy each cell controller is viewed as a VMD by the area controller. Later we shall see that this allows the same integration and ME aggregation tools to be used to implement their distinct manufacturing tasks.

In order to enable the various manufacturing entities to interact, an appropriate communications architecture was chosen and implemented. For the experimental conditions described, a flat network architecture was chosen to enhance the flexibility of the overall system. MAP (Manufacturing Automation Protocol) was selected because it conforms to the OSI seven-layer model (Zimmerman, 1980), is vendor independent, is suitable for use in manufacturing shop-floor environments and is available for a range of computing hardware. Each manufacturing entity (with its particular functionality and communications protocol) had to be connected to this communications architecture. In order to 'convert' these various MEs into VMDs (see discussion in section 2.2), configurable gateways were designed and built. These provided protocol conversion and selectable local processing facilities (Weston *et al.*, 1985).

Both the Adept vision system and Adept robot were equipped with integral controllers. These, however, did not support MAP. Therefore an additional gateway was required to establish a link between them and the MAP network. The Adept controllers support RS232 serial communications, so each Adept controller was connected via such a data link to an IBM AT personal computer which incorporated a MAP interface in its PC Bus backplane (see Figure 2.5(a)).

The communications protocol used between the Adept applications and the IBM gateway, part of which had to be included and run within the Adept controller, was specified and implemented in house. This protocol was based on a four-layer model which maintained data integrity across the RS232 serial link (see Figure 2.5(b)). Concord Communications Inc. MAP 2.1 hardware and software were obtained for the IBM PC AT. Proprietary software, which provided a simplified user interface to MAP CASE (Common Application Service Elements, 1988), was also purchased and installed. This provided the interface between the MAP communications protocols and the gateway. In addition, a number of debugging modules were added to aid in monitoring information flow via the gateway and to enable the gateway to generate messages locally and issue them to connected manufacturing entities.

The assembly cell workpiece transport system comprised a Bosch conveyor, an in-house-designed control unit and control software running on an IBM personal computer. The controller architecture used was the subject of ongoing research in the design of distributed hierarchical control systems for materials handling. The conveyor controller, however, was a standalone device and so had to be interfaced to the MAP network. MAP hardware and software identical to that used for the Adept gateway were installed, and hence the MAP interface module constructed for the gateway could be used as the basis for the conveyor controller MAP interface.

The palletizing/loading station was to be operated manually, thus requiring an interface between the human operator and the cell controller. In order to implement this interface without additional MAP hardware and software, the conveyor controller's screen and keyboard facilities were utilized. Messages could then be exchanged via the conveyor controller's MAP interface and routed separately to an operator window.

The choice of computer hardware and software to perform cell control tasks was constrained by a number of criteria. The computers needed to support MAP in order to communicate with other manufacturing entities. A real-time multi-tasking operating system was also required due to the variety of tasks to be performed by the cell controller. An Intel 310 microcomputer with Industrial Networking Inc. (INI) MAP hardware and Intel MAPNET software was selected; the operating system used was iRMX86 which provided the necessary run-time support. As stated earlier, the required system also included the management of two such cell control sites by an area controller. Here, the choice of machine was a SUN 3/160 workstation equipped with INI MAP hardware and Sunlink OSI software.

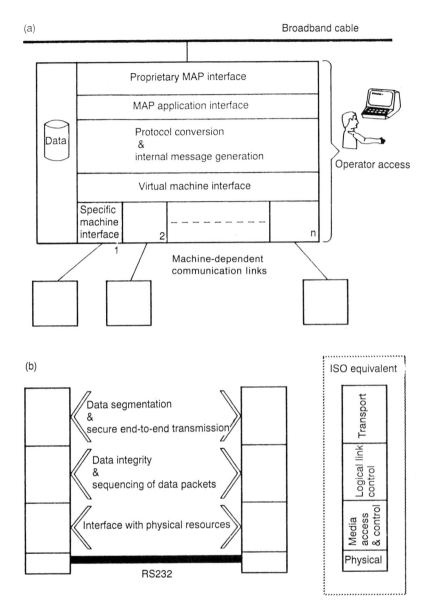

Figure 2.5 *(a) Adept gateway, (b) four-layer protocol*

In the above system, because of the unavailability of MAP MMS, its predecessor MMFS (Manufacturing Message Format Standard) (Zimmerman, 1980) was used. Its relative simplicity allowed MMFS conformant code to be written by the SI research group where it was not available commercially. This has involved the authors in having to consider methods of making manufacturing devices appear to be standard (or virtual) ones.

The required functionality of the various manufacturing entities and their interrelationships then had to be established and the information flow between them specified. The functional relationships so established were to be implemented in such a way that they could easily be redefined and reimplemented. Each of the four machine-level manufacturing entities was to perform a well-defined manufacturing task. In the case of the vision and robot systems these tasks needed to be implemented using the VAL II programming language. Owing to the fundamentally different architecture of the conveyor controller, its manufacturing tasks could be decomposed into a collection of primitive tasks, each of these to be managed by the cell controller.

3.3   AUTOMAIL: A Configuration Tool

It is clear that the cell controller must be capable at run time of managing the interactions between the machine-level devices under its charge. As stated earlier, these interactions could be 'hard wired' (albeit in software) in the sense that they can be specifically described by some operational specification. If the specification changes with time, which is very likely (e.g. new machines are added, different operator practices are required, new products are substituted or added, different higher-level computer systems are introduced, etc.), changes will not easily be made if conventional practices of implementing to a fixed specification are adopted.

Furthermore, as previously indicated, cell controller vendors and systems integrators will have to deal with many types of system configuration and ME. Thus, high-cost specific customized solutions will have to be engineered to meet individual requirements and will seldom be justified on an economic basis.

A set of configuration tools which complements the 'run-time' elements is clearly required. Such tools can build on the VMD principle where a consistent approach to defining and debugging VMD interactions can be realized in the form of an 'integration language'. This can be widely applied to reduce systems engineering when installing or modifying systems. The above requirement was identified by the System Integration Group at LUT. The outcome was the AUTOMAtion Integration Language (AUTOMAIL) (Sumpter et al., 1987) which comprises a distributed system application language and a set of configuration management, concurrent task management and debugging tools (see Figure 2.6).

For general cell control applications, AUTOMAIL supports the programming and management of MEs on an individual basis, that is decomposed program tasks can be defined with the degree of granularity required to specify and control ME task and intertask interactions in a cell. This approach facilitates reconfigurability as ME control tasks can readily be replaced, inserted or modified in response to changing requirements. AUTOMAIL not only supports the development of individual ME tasks

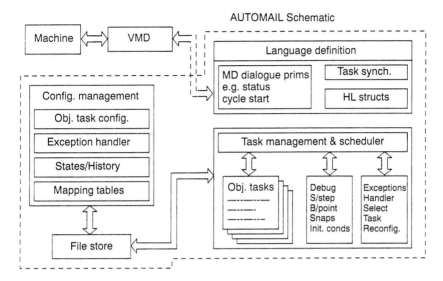

Figure 2.6 *AUTOMAIL schematic*

but, at run time, supports the execution of these tasks in a multi-tasking asynchronous manner by providing task interaction mechanisms based on communication primitives which use the communication service. The control tasks themselves have a substructure based on the concept of a 'function module' which typically contains a small number of program statements, grouped together to initiate an action or to attain a particular program state.

It is important to appreciate that the AUTOMAIL methodology applies not only to conventional shop-floor MEs but also to software MEs, which could reside within the cell controller (such as a 'real-time' scheduling agent), in a configurable gateway (such as a 'production-monitoring' agent) or at higher-level entities (such as a 'bidding' agent in a shop controller). In each case the VMD principle can be applied so that a consistent approach to interaction issues can be taken.

An extract from typical decomposed AUTOMAIL function module definitions is shown in Figure 2.7. This illustrates how concurrently operating tasks, relating to MEs of the example cell, can be defined and synchronized.

Exception handling in AUTOMAIL is programmed away from the main control task flow. If an exception occurs during the execution of any function module, subsequent program flow is directed through user-written handlers. Here, control task sequencing can be modified, for example individual function modules can be replaced and reorganized, or complete control tasks replaced. This approach facilitates, for example, orderly shutdown or cell rescheduling on a dynamic basis. Task synchronization (e.g. the halting of one manufacturing process until another completes or

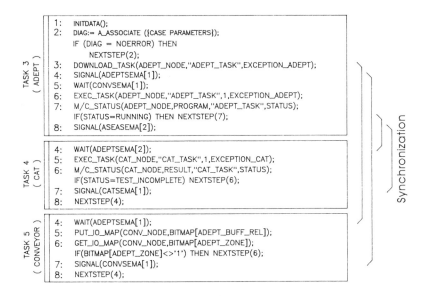

Figure 2.7   *AUTOMAIL – extract from task definition*

reaches a desired state, such as with a pallet transfer) is also supported by standard program semaphore mechanisms.

The shop controller is directly responsible at run time for interacting cell controllers. Here, the AUTOMAIL methodology has also been used with a parent AUTOMAIL in the shop controller which is used to define, debug and run child AUTOMAILs in each cell controller, again utilizing the VMD concept. In this way, interactions at all hierarchical levels can be flexibly defined and managed, thereby minimizing the need for bespoke applications.

To illustrate the granularity with which decomposed ME tasks can be defined and synchronized, two uses of the AUTOMAIL methodology from the previously described manufacturing scenario will be detailed further. The workpiece transport system was segmented into a number of zones and an AUTOMAIL task executed to interrogate the conveyor controller to obtain the relevant positions of all pallets on the conveyor. Through processing information, the AUTOMAIL task can detect the presence (or absence) of a pallet in an associated zone and initiate relevant manufacturing operations. This illustrates the relatively high-speed processing of real-world states within the cell controller, involving electronic transfer of those states via the communication service. An alternative example use of AUTOMAIL is illustrated in Figure 2.7, where a named VAL II program is loaded via the communications architecture from disk storage on the cell controller to the AdeptOne robot, thereby transferring large files of information. Subsequent AUTOMAIL task statements will be executed to generate software state changes, to initiate the running of that program in the robot controller and to take appropriate action on completion of the assembly task so invoked.

Many additional AUTOMAIL tasks have been defined and run using the cell control architecture described. These tasks have facilitated:

(1) interpretation, decomposition and distribution of work-to-lists generated by a proprietary production planning and control (PP&C) package and routed via the shop controller to child AUTOMAIL tasks at the cell level;

(2) provision of decision support functions, via terminal access, to facilitate manual decision making at both shop and cell levels;

(3) access to database information relevant to each hierarchical level via SQL (Standard Query Language) update and retrieval mechanisms (ANSI/SQL, 1986), this work centring on access to both PP&C (Rui, 1989) and CAD/CAM (Chan, 1989) simulation and design MEs.

The above-described cell control architecture remains as a test bed for the evolution of integration approaches and demonstrating their effectiveness. The facility for decomposing a manufacturing task into a number of separate AUTOMAIL tasks (to be executed asynchronously by the cell controller), and to define the interactions between tasks based on communication primitives, has demonstrated fully an ability to achieve integration on an incremental basis. This is an essential requirement when creating complex factory-wide integration schemes.

The SI group's AUTOMAIL can be viewed as a 'cell control shell' as it provides an underlying resource which can build, debug and run concurrently operating ME task descriptions with much reduced systems engineering cost and time. Essentially the shell represents a set of tools which has been designed to meet the generic needs of cell controllers. However, application-specific software will always need to be created to achieve the specific functionality required. The AUTOMAIL methodology significantly minimizes this exercise by providing tools and constructs which generate expandable and supportable solutions.

The studies described have also highlighted major problem areas for which solutions must be derived before the full benefits of CIM approaches can be realized. In particular, the difficulties of accomplishing interaction with proprietary software packages run in a heterogeneous distributed environment (such as PP&C and CAD/CAM systems) have identified the need to reconsider the way in which such MEs should be designed (Hodgson *et al.*, 1988). Similar observations can be made with respect to machine-level MEs (Weston and Hodgson, 1989). In particular the SI group has identified the definitive need to separate out action causing functions (commonly of a procedural nature) from information (Weston *et al.*, 1989c). We shall return in section 5 to the reasoning behind this statement. Through such a separation major advances can accrue with respect to configurability, potentially offering a way forward in accomplishing genuine information-driven product realization.

4   GENERAL OBSERVATIONS RELATING TO CELL CONTROL IN A CIM ENVIRONMENT

In the previous sections we have seen that MEs represent a decomposition of a complete manufacturing system into functional units which can be treated separately. Such a decomposition can result in an ability to segment any problem into containable subproblems and can form the basis of a standardization methodology. Subsequently, however, there will be the need to aggregate (or reassemble) MEs to accomplish specific manufacturing tasks. This aggregation will be by no means trivial but can be achieved through the use of 'information-driven' techniques on the assumption that all MEs comprise, at least in part, a computer system. As previously stated, there are significant differences between the various possible MEs, but we have also observed an important similarity in that they all function to change states (be those software-encoded information or real-world states).

Figure 2.8 illustrates an internal view of a cell controller, depicting a feasible decomposition into more primitive MEs. It should be remembered, however, that each of those decomposed MEs could themselves comprise even more primitive MEs. For example, the planning entity could be decomposed into an 'interpretation ME' which interprets higher-level planning information, a 'bidding ME' which might negotiate with external entities for work, and a 'task generation ME', which could use some manually driven, algorithmic or inference method of generating synchronization and sequencing information (i.e. lower-level plans) for lower-level entities. It is unlikely that any set of rules can be consistently applied in decomposing such problems, but the purpose of modularizing should not be forgotten – that the problem is simplified, and/or that an entity already exists for the purpose, and/or that a standard or at least consistent approach to problem solving can be achieved. Table 2.2, which should be referenced in conjunction with Figure 2.8, illustrates the purpose of the other possible internal entities of the cell controller. The relative importance and complexity of these primitive building elements will depend upon the type and purpose of the shop-floor functions to be managed. If, for example, interaction between a number of dissimilar machines is involved, it may be that synchronization and sequencing issues are the most difficult to address, whereas with a number of similar machines, planning issues (including scheduling) may require a sophisticated solution.

It can be argued that MEs already exist, or are being derived through manufacturing systems research worldwide, to provide cell control functionality of the type depicted in Figure 2.8. However, like machine-level

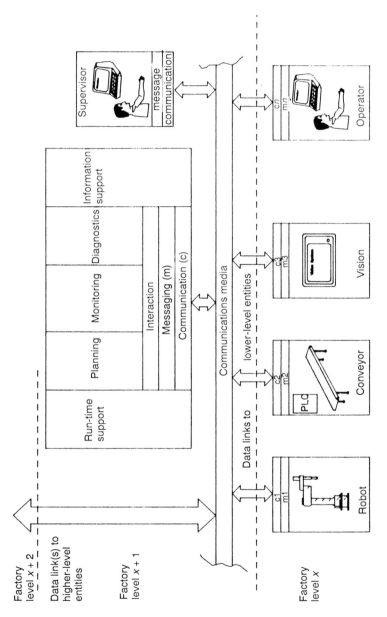

Figure 2.8 *Internal view of a cell controller*

Table 2.2   *An internal view of cell control entities*

Communication	Data exchange
Messaging	Information exchange
Interaction	Synchronization
	Sequencing
Information support	Information access & storage
	Configuration data
Run-time support	Operating system interface
	Exception handling
Planning	Multi-task interpretation
	Multi-task generation
	Bidding mechanisms
Monitoring	Productivity information
	Quality information
	Traceability data
Diagnostics	Maintenance reporting
	Condition monitoring

MEs generally, they have been designed in a standalone manner, often providing an elegant solution to only a part of the overall problem of 'realizing products more efficiently than the competition'. Thus, almost invariably, existing cell controllers represent solutions to parochial or tactical problems (such as how can real-time scheduling of ten machines be achieved?) rather than more global or strategic problems (like how can improvements in 'time to market', product 'lead times', quality and profitability be realized?). Clearly this same problem, of utilizing existing or emerging MEs which have not been designed to facilitate integration, exists at all levels in the factory hierarchy.

We have seen that the VMD concept is evolving to address this problem. However, until a VMD specification for all types of manufacturing system component has been fully derived and widely agreed upon, the designers of individual components cannot work in sympathy with system builders to meet users' overall requirements. A temporary approach to this problem (which unfortunately may be required for some time to come) is that adopted by the SI Group and described in section 3 where external processing facilities are 'attached' to each ME so that it can be viewed for integration purposes as a standard building component. In this way, as previously explained, VMDs can be disaggregated or subsequently re-aggregated, in a generic manner. Ultimately, it will be the responsibility of the suppliers of individual MEs to provide this processing facility, but this situation cannot fully be reached until further research is completed in specifying integration frameworks and the resulting integration requirements of VMDs.

5   INTEGRATION ARCHITECTURES AND THE ROLE OF AUTOMAIL

Throughout this chapter the danger of creating very-large-scale inflexible integration schemes has been stressed. Even if formal modelling methods (for example, derivatives of CIM OSA (Weston *et al.*, 1989c)) are used to reflect strategic business goals and to map these goals efficiently into optimized tactical solutions which can be implemented effectively, they will not remain as optimal solutions for long in typical manufacturing domains.

The AUTOMAIL concept was not primarily conceived to create individual manufacturing applications, but to complement formal modelling methods by providing an implementation tool for aggregating MEs in a highly flexible and efficient manner. Only where MEs do not already exist, or cannot be moulded into the form of a VMD, are AUTOMAIL-derived application processes created (e.g. a simple scheduling heuristic).

However, as various application scenarios have been studied and their generalized features extracted, the scope of AUTOMAIL has been widened considerably to cope with the range of possible MEs and the difficulties of treating them in a standard way. These difficulties are particularly evident when establishing links to proprietary PP&C and CAD/CAM simulation and design packages. Like most MEs, such packages are typically an internally integrated bespoke mix of (1) action-causing (usually procedural in nature), (2) information processing and, occasionally but increasingly commonly, (3) communication software functions. However, it has become apparent to many researchers that major advantages can accrue from separating out these three classes of function. In particular much improved implementation independence, configurability and maintainability can result. In addition, information technology (IT) developments in the areas of high-level languages, database design and communications protocol have provided the tools to realize those advantages in a cost-effective manner. When considering factory integration schemes and ME interactions the same emerging design rules should be applied, particularly as heterogeneous multi-vendor implementations will be the norm.

The SI Group's integration studies have identified the need for three separate integration architectures (as depicted by Figure 2.9) to enable highly configurable, visible and maintainable CIM systems to be created (Weston *et al.*, 1989b). Each architecture can provide an essentially transparent set of services to MEs on a factory-wide basis and itself be served by configuration management tools to enable information-driven integration. A study of the necessary features of these three architectures

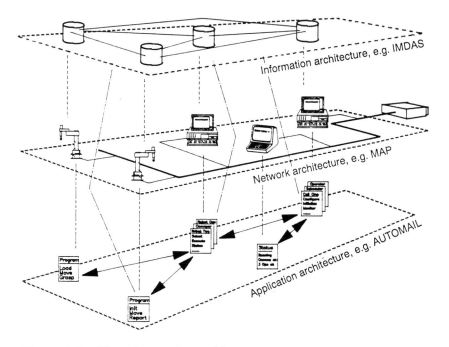

Figure 2.9   *Three integration architectures*

and the resulting implications on VMD specifications is the subject of ongoing work of the group. As a means of evaluating the concepts involved, the systems integration tools derived are being developed further and installed at a manufacturing plant of a major UK computer vendor. The features of these implementation studies will be considered in the next section, but first the necessary features of the three integration architectures will be briefly described.

5.1   Network Architecture: Communication Services

The function of the network architecture is to enable the meaningful exchange of information between MEs, which will normally demonstrate heterogeneity. The architecture therefore provides a range of services which should be accessed via standard call mechanisms. Information transfer will often be over various point to point and LAN data links and may even involve WANs (wide area networks) where integration of more than one company is involved. The network architecture would therefore ensure that message transactions initiated by MEs are channelled via the correct communications route (e.g. network drivers which might reside in separate MAP and TOP protocol stacks), thereby satisfying the requirement for a virtual connection. Through the provision of standard access mechanisms to a separate architecture, the manufacturing application is decoupled from the details of how information is transferred. This means that the manufacturing application and the network services can be developed independently, which can have very significant benefits when changes are required.

5.2   Information Architecture: A Distributed Library of Information

The IT community has taken major steps forward in producing database tools during the last decade. More recently much of this work has centred on the design and creation of distributed databases, resulting in products which are now widely applied in banking and many other service industries, often on a worldwide basis. However, there remain major difficulties which must be overcome before distributed database technology can be consistently applied at all levels of the factory hierarchy. In particular very large quantities of information can be involved, where the necessary rate of updating and retrieving information at heterogeneous MEs may often present a requirement which is beyond the capabilities of contemporary systems (Chu, 1988). Additionally, much work remains in specifying product models which can be widely adopted at each level in different manufacturing domains (PDES, 1989).

Although distributed database technology is an emerging science, its predecessors, centralized databases, have been used for many years within manufacturing organizations. However, their use has been largely confined to enterprise and factory levels of the manufacturing hierarchy, dealing separately with information services relating to finance, marketing, engineering design and issuing, production control and quality. Often large investments have been involved in installing and generating information stored in such systems where information update, retrieval and management services are manually driven. It is typically untenable to consider the wholesale replacement of such systems when aiming to achieve electronic interaction with other MEs (such as cell controllers). Therefore, interim solutions which provide external processing may be required to realize, at least in part, the benefits of a VMD approach. In the not-too-distant future, however, the increasingly widespread adoption of standard query languages (such as ANSI SQL) and information modelling and schema management tools (Rui, 1989) will enable the extension of emerging distributed database concepts to be used at least at the shop and cell levels of the hierarchy. This must, however, be complemented by a major change in design methodology at individual MEs, where information services need to be separated out to realize the full benefits of these IT tools.

The implications of this discussion are that a further decomposition is highly desirable so that the problems of managing, updating and retrieving information can be delegated to a set of information services which can deal transparently with information modelling, accuracy, integrity and security issues. Again, the manufacturing applications can be decoupled from complex information modelling and management issues, allowing for example the potential for independent modification of product descriptions (as product variants are introduced) and production methods (which will be more procedural in nature). Amongst other potential advantages, the impact of such an approach will be to reduce significantly the 'time to market' of new products and provide a graceful migration of production techniques.

As is the case for the network architecture, the information architecture should provide standard call mechanisms (such as SQL) and itself be served by configuration management tools to minimize systems engineering cost/time.

5.3   The Application Architecture: Dealing with Manufacturing Actions

The earlier description of AUTOMAIL highlights a methodology for building, debugging and running manufacturing applications, thereby determining the way in which MEs (including people) interact and perform actions to realize product manufacture. As previously stated, the type and role of MEs will be specific to a particular manufacturing organization so that some bespoke creation of ME task descriptions will be inevitable. However, the AUTOMAIL methodology can minimize considerably the implementation problems in this area through use of its 'shell of services'. Current versions of AUTOMAIL already utilize underlying network architectural services, whilst ongoing work is aimed at realizing generic access to information architectural services which themselves are based on distributed database technology. Figure 2.10 illustrates the role of the three integration architectures.

This figure shows how manufacturing applications on a single machine in an integrated system are insulated from each other and the network services. The external flexibility is enabled by the services related to the three architectures. Each of these is functionally independent of the others so that development, enhancement or varying degrees of implementation may occur.

The internal flexibility arises from the methods and tools used in the construction and development of the applications themselves.

It should be noted that the sophistication of an implementation to this plan will vary widely. Thus, an AUTOMAIL cell controller might contain high levels of information, application and network services. On the other hand, an AUTOMAIL-based machine controller might have very low levels but still appear as an AUTOMAIL machine when viewed 'down its network connection'. Furthermore, the network services will contain interfaces to machines which are not AUTOMAIL based at all. In this translation of AUTOMAIL transactions will take place at this level.

By these means systems of any size can be built containing almost

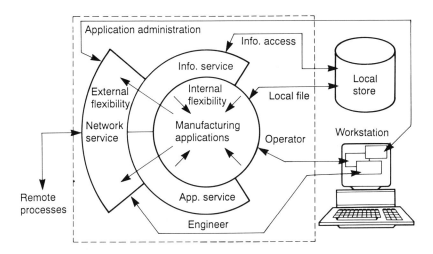

Figure 2.10 *Role of three integration architectures*

completely existing system components and according to the current perceptions and practices of the implementors yet still offer the necessary flexibility for future system change.

6   MAIN FEATURES OF AN INDUSTRIAL CELL CONTROL DEMONSTRATOR

The cell control philosophies previously described in this chapter are currently being assessed and enhanced through industrial implementation studies at a UK manufacturing plant. This company designs and manufactures a wide range of complex, multi-layered, mixed technology printed circuit boards in small to medium batches for its various mainframe computers, personal computers and associated system components. The company has been actively involved in systems integration work for a number of years and has formulated a wide-ranging policy in the area aimed at significantly increasing its competitive edge in this particularly dynamic and cut-throat market sector. As part of its evolving integration methods, the AUTOMAIL concept is initially being assessed in the assembly and test areas with a view to creating an integration platform on which reconfigurable cell and shop controllers can be produced. The inherent flexibility of AUTOMAIL enables a consistent approach to factory integration which, on a wider front, can lead to the creation of highly efficient integration schemes. These have the capacity to facilitate change as the market and other factors dictate.

Ongoing implementation work at the manufacturing plant concerns the specification and installation of cell controllers in the automatic component

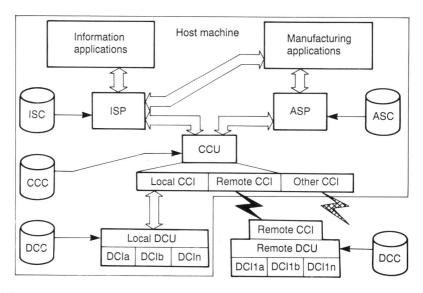

Figure 2.11 *AUTOMAIL platform*

insertion (ACI) and surface-mounted (SM) component assembly areas. The
AUTOMAIL platform for this work is depicted by Figure 2.11. Again, it
should be stressed that the VMD concept can only be adopted through the
provision of 'front-end' processing for existing MEs, as the activities of a
wide range of these need to be integrated. In this case they include
automated machinery, operator/engineering decision support facilities,
product issuing and archiving systems, quality systems and planning and
control agents. The evolving integration schemes will be highly flexible,
ensuring the following.

(1) *Manufacturing strategy independence.* No part of the integration
    infrastructure is inextricably linked with the particular
    manufacturing strategy or tactics involved, for example
    JIT, MRP, OPT, etc. It will, therefore, be possible to
    select/change/configure these approaches afterwards.

(2) *Manufacturing application portability.* The methodology re-
    duces as far as possible the dependence of an application on its
    local environment in terms of its interfaces to, and interactions
    with, other applications and information stores.

(3) *Individual device implementation.* The adoption of the scheme
    does not entail a large amount of overhead processing ability by
    a given ME.

(4) *Multi-device implementation.* It is not practical, or even desir-
    able, for all parts of the system with which a subpart must
    interact to be changed simultaneously. The methodology offers
    a stepwise method of introducing integration, where sub-
    systems can be proven on an incremental basis.

7   GENERAL OBSERVATIONS

This chapter has highlighted the problems of simply specifying solutions to cell control problems (and more generally integration problems), then implementing those solutions in an *ad hoc* way. The result almost inevitably will be non-ideal and subsequently will be viewed as a larger hard-wired inflexible machine which cannot cope with changing requirements nor draw on advancing enabling technologies.

Furthermore such a bespoke approach will lead to duplicated effort worldwide, the cost of which inevitably will fall on the industrial user. Better by far is an approach based on the adoption of an integration framework and system-building tools.

Much work has been done in facilitating the latter approach although clearly significant work remains, as does the need to convince vendors and users alike of the potential benefits. This chapter has shown how interim system-building methodologies and tools can be used with significant benefit, particularly in industrial environments where an installed base of manufacturing systems already exists and must be utilized.

As a final comment, it is emphasized that cell controllers, and systems integration schemes more generally, are not synonymous with mechanized automation and/or very high levels of equipment investment. Integrating the activities of people is equally important, which leads on to the conclusion that CIM is not out of the reach of small- and medium-sized enterprises. It is important to encourage this view if efficient supplier, manufacturer, subcontractor and customer relationships are to be formed.

ACKNOWLEDGEMENTS

The authors wish to acknowledge fully the contributions of C. M. Sumpter, a former SI Group member who was responsible for initiating much of the AUTOMAIL methodology and has since formed his own consultancy company. In addition, the research studies have been funded by the Application of Computers in Manufacturing Engineering (ACME) Directorate of the UK Science and Engineering Research Council, with significant input made from technical staff of International Computers Ltd (ICL) at Kidsgrove, England.

REFERENCES

ANSI/SQL (1986). *ANSI American National Standard Database Language SQL.* New York: American Standards Institute Inc.

Chan, X. (1989). Advancements in robot programming with specific reference to graphical methods. *PhD Thesis*, Loughborough University of Technology.

Chu, W., ed. (1988). *Distributed Systems, vol. 11: Distributed Data Base Systems.* Deham, MA: Artech House.

Common Application Service Elements Specification (1988). Part of *MAP Version 3.0 Specification*. Society of Manufacturing Engineers, Detroit, MI, USA.

Hodgson, A. *et al.* (1988). Planning and control information flow in CIM, current research directions and the need for intermediate solutions. *IERE Int. Conf. on Factory 2000, Cambridge.*

Klittich, M. (1988). CIM-OSA, the implementation view point. *Proc. 4th CIM Europe Conf., Kempston.* Bedford: IFS and Berlin: Springer.

Manufacturing Message Services Specification, Part 1: Service Specification, Appendix B: Guidelines for Writing Companion Standards (1987). EIA Project 1393A Draft 6.

MAP and TOP Version 3.0 Specifications (1988). Society of Manufacturing Engineers, Detroit, MI, USA.

OTTAWA (1986). *Reference Models for Manufacturing Standards.* Internal Report N51, ISO 184/SC 5WG1.

PDES (1989). *A Reporting of the PDES Initiation Activities.* Bradford M. Smith, Room A353, Bldg 220, National Bureau of Standards, Gaithersburg, MD 20899, USA.

Rui, A. (1989). Steps towards computerised administration of factory information resources in CIM. *PhD Thesis,* Loughborough University of Technology.

Sumpter, C. M., Weston, R. H. and Gascoigne, J. D. (1987). Computer integration in flexible assembly systems. *Int. J. Rob. Res.,* **10**, 43–9.

Van Dyke Paranuk, H. and White, J. F. (1989) *Synthesis of Factory Reference Models.* ITI TR 87-29 (Industrial Technology Institute, PO Box 1485, Ann Arbor, MI, USA).

Weston, R. H. and Hodgson, A. (1989). *Systems Integration, ACME Final Report.* Grant GR/D46984, Science and Engineering Research Council, Swindon, UK.

Weston, R. H., Gascoigne, J. D. and Sumpter, C. M. (1985). Intelligent interfaces for robots. *IEE Proc. D, Control Theory and Applications,* July, 168–73.

Weston, R. H. *et al.* (1988). Steps towards information integration in manufacturing. *Int. J. CIM,* **1**, no 3, 140–53.

Weston, R. H., Gascoigne, J. D., Sumpter, C. M. and Hodgson, A. (1989a). Robot integration with CIM. *Int. J. Prod. Res.,* **27**, no 3, 513–28.

Weston, R. H. *et al.* (1989b). Configuration methods and tools for manufacturing systems integration. *Int. J. CIM,* **2**, no 2, 77–85.

Weston, R. H., Gascoigne, J. D., Sumpter, C. M. and Hodgson, A. (1989c). The need for a generic framework for systems integration. In *Advanced Information Technologies for Industrial Material Flow Systems* (Nof, S. Y., Moodie, C. L., eds), NATO ASI Series F53, Springer, pp. 279–309.

Zimmerman, H. (1980). OSI reference model – the OSI model of architecture for open systems integration. *IEEE Trans. Commun.,* **TC28**, 425–32.

3 *The cell as part of a manufacturing system*

Albert Jones and Abdol Saleh

1   INTRODUCTION

In this chapter, we describe a new approach to the design, implementation, and integration of cell controllers in a manufacturing system. It combines techniques from control theory, operations research, and computer science. This cell controller can be (1) modified to fit both the physical definition of a cell and the capabilities of other controllers in the system, and (2) easily integrated into any shop-floor control system which meets the interface requirements.

1.1   The Automated Manufacturing Research Facility

The results described in this chapter are based on the experience gained in building the Automated Manufacturing Research Facility (AMRF) (Simpson *et al.*, 1982) at the National Institute of Standards and Technology in the USA. The AMRF is a prototype small batch manu-facturing facility built to address two important issues: integration stan-dards and measurement techniques in an automated factory. Physically, the AMRF contains a variety of robots, machining centres, a coordinate-measuring machine, and an automated guided vehicle. This equipment has been integrated together using three separate architectures: shop-floor control, data management, and communications (see Figure 3.1).

The AMRF shop-floor control architecture is a four-level hierarchical system (Jones and McLean, 1986) (see Figure 3.2). Currently, only the bottom three levels have been implemented. Each piece of equipment has its own AMRF-built controller. This equipment is grouped into small units called workstations. Each workstation is designed to perform a specific activity (e.g. milling, turning, inspection, material transfers, etc.) and has its own controller. The workstations are managed by the cell controller. All planning and scheduling decisions are made at cell level. Each workstation receives one instruction at a time. It first decomposes that instruction into the tasks to be performed by the equipment under its control and then coordinates the activities of the equipment as required. The task decompo-sition is, in general, completely deterministic with little or no flexibility.

AMRF researchers have designed and implemented an architecture called IMDAS (the Integrated Manufacturing Data Administration Sys-tem) to manage all data (Libes and Barkmeyer, 1988). IMDAS has been

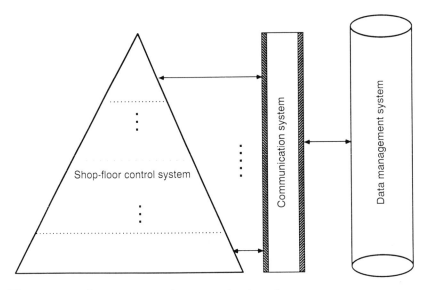

Figure 3.1 *Separate control, communication, data management systems*

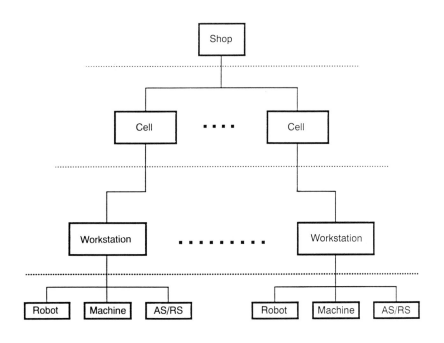

Figure 3.2 *AMRF shop-floor control architecture*

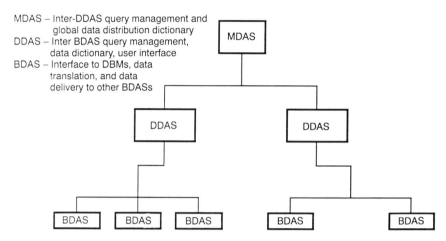

Figure 3.3 *IMDAS architecture*

specifically designed to operate in a distributed, heterogeneous computing environment in which (1) control computers have time-critical data needs, and (2) data resides in a variety of commercial databases. IMDAS is completely separate from the control hierarchy and transparent to the modules in that hierarchy (its users). Users simply request data from IMDAS in a standard way. IMDAS then retrieves the data wherever it is and provides it to users in the format they desire. Users are totally unaware of the effort required to answer their requests. The AMRF view is that IMDAS plays the same role in managing its resources (data and data repositories) that the shop-floor control hierarchy plays in managing its resources (material and equipment). Hence, IMDAS is a three-level hierarchy of data management services: the basic (BDAS), the distributed (DDAS), and the master (MDAS) data administration service modules (see Figure 3.3). Each BDAS can be tied to multiple data repositories. Detailed descriptions of these functions can be found in Liber and Barkmeyer (1988).

In the AMRF, processes communicate with each other by writing and reading messages in memory areas that are accessible by both the process and the communications system. These 'common' memory areas are called mailboxes. The network communications system is responsible for delivering messages from the source mailboxes, written by applications processes, to any destination mailboxes that are logically connected to them. For those processes residing on the same computer system, this is very simple, but for those processes residing on different computer systems, an external network is required. The AMRF network architecture contains several subnetworks

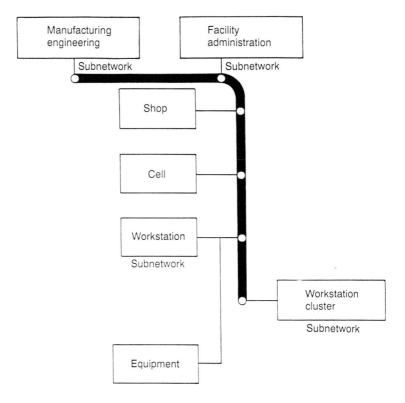

Figure 3.4   *AMRF network architecture*

linked to a large backbone network (see Figure 3.4). Each workstation has its own local area network linking its equipment controllers. These subnets are either RS232 or Ethernet and ensure a quick response for time-critical operations. The backbone network will be a broadband, token-bus network. Details about the evolution of the AMRF network can be found in Rybczynski *et al.* (1988).

We will return to the AMRF later in this chapter.

1.2   Chapter Overview

In section 2 we describe a cell. In section 3, we examine the classical approaches to controlling dynamic systems and review some of the present applications in manufacturing. Section 4 contains a description of the two most popular approaches to shop-floor control: hierarchical and heterarchical. In section 5 we detail our approach to designing a cell controller and discuss information requirements and implementation issues. We also provide a summary and bibliography.

2   WHAT IS A CELL?

Currently, there is no standard, or even accepted, definition of a cell. In industry, one finds two types of cells: they contain either a collection of identical, or functionally identical, equipment or they are group technology cells. In the first case, a cell might contain a collection of drills, or a collection of milling machines, etc. In the second case, a cell might contain all of the equipment (sometimes a single piece) needed to manufacture a particular family of parts, assemble a specific family of circuit boards, etc.

The AMRF introduced a totally different kind of cell. The AMRF cell contains physical groupings of equipment called workstations (see Figure 3.5). There are four types: machining, cleaning/deburring, inspection, and material storage/transportation. The architecture described below can be used for each of these 'cells'.

There are many companies marketing 'cell' controllers even though there is no standard definition of a 'cell'. In addition, there are no standard internal functions, external interfaces, or hardware platforms. There are two classes of vendors, and two different design philosophies. Vendors of equipment controllers have based their cell controller designs on their existing programmable logic controllers (PLCs). Those extensions provide the communication and database access necessary to interface with and coordinate several lower-level PLCs. Many of them even provide limited scheduling. The controllers marketed by 'system integrators' typically contain sophisticated scheduling and database capabilities; some even use expert systems.

The major problem with this scenario, at least from the users' perspective, is that it is virtually impossible to determine what type of cell controller to buy and how to integrate it into an existing or future manufacturing system. The architecture presented in this chapter also addresses this issue.

Before we describe our cell controller, we discuss two important background topics: decentralized control of dynamic systems and shop-floor control systems.

3   DECENTRALIZED CONTROL OF DYNAMIC SYSTEMS

Sandell *et al.* (1978) distinguish between two types of decentralized methodologies to control the evolution of dynamic systems: multi-layer and multi-level. Multi-layer controllers deal with the fact that decisions are made and events occur at different frequencies in the same system. However, these types of controllers typically do not specify how decisions are related to one another or how events influence those decisions. Multi-level control, on the other hand, provides a methodology for decomposing complex decisions into smaller, simpler ones, and, in certain cases, solving

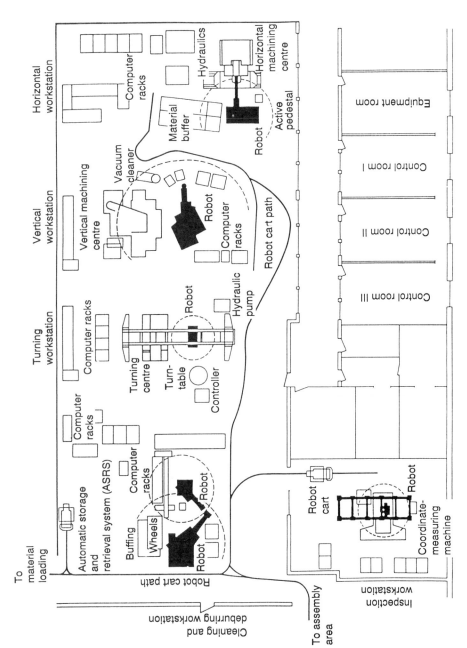

Figure 3.5 *AMRF shop-floor layout*

them to optimality. There is, in general, no dependence on time or frequency but, as we will see, it can be used to model the relationships between different decision at the same frequency and the same decision at different frequencies.

3.1   Multi-layer control

Important events which influence the behaviour of large, complex, interconnected systems typically occur at different timescales. Modelling these systems often begins by defining state variables and state transition functions for these events. Events which occur at the same, or nearly the same, frequency can be clustered into groups. These groups form the layers of a multi-layer controller. Each layer operates on a different 'timescale' and uses different sets of aggregated information. The assumption is that state variables *within* a particular layer have 'strong' interactions and those *across* layers have only 'weak' interactions. More details on the mathematical models and structure of these controllers can be found in Jamshidi (1983).

Albus (1981) and Saridis (1985) pioneered the use of multi-layer control in robotics and both used a three-layer model. In the Albus model those layers are called TASK, E-MOVE, and PRIMITIVE. The TASK layer determines a plan, a series of robot moves for executing each new robot command. The E-MOVE layer determines an optimal trajectory for each of these moves. The PRIMITIVE layer provides the interface to the robot and monitors the execution of each move. In the Saridis model the three layers are called ORGANIZATION, COORDINATION, and EXECUTION. They perform very similar functions to those found in the Albus model. The ORGANIZATION layer does planning, the COORDINATION layer chooses actions to carry out the plan, and the EXECUTION layer interfaces to a specific robot.

Gershwin (1989) recently used the notion of multiple timescales to propose a mathematical justification for hierarchical analysis of production systems. The formulation of the decisions in this system contains both continuous and discrete variables. Furthermore, those decisions can contain deterministic, stochastic, linear, or non-linear terms. Gershwin used the frequency separation methodology discussed above to propose his hierarchy. He placed events that occur very infrequently at higher layers and those that happen very frequently at lower levels. The mathematical relationships needed to control events at higher layers ignored the details of the variations of the events occurring at the lower layers. The formulations at the lower layers viewed the events at the higher layers as static, discrete events.

Villa and Rossetta (1986) addressed the temporal relationships that exist between the layers inside a multi-layer controller. They proposed that each controller have three planning parameters: a planning horizon H, an updating period P, and a sampling period, T. The authors argued that a controller will perform efficiently if $H > 10 \times P$ and $P > 10 \times T$. They also indicated that it was possible to have more than one multi-layer controller superimposed on top of one another. They proposed that the interfacing between controllers in adjacent levels could be achieved by setting a lower level's planning horizon and updating period to the upper level's updating and sampling periods, respectively.

3.2   Multi-level control

Mesarovic *et al.* (1970) presented one of the earliest, formal, quantitative treatments of multi-level control systems. The techniques for problem decomposition are based on methods from the theory of decomposition for mathematical programming problems (Geoffrion, 1970). The purpose is to decompose a complex optimization problem into a series of smaller and simpler subproblems. Most decomposition procedures result in a two-level structure (see Figure 3.6) and use conditioning of either the objective function (Dantzig and Wolfe, 1960) or the constraint set (Benders, 1960).

Hax and Meal (1975) were one of the first to apply these concepts to a production planning problem. It is important to note that their decomposition of the production planning problem led to a similar aggregation/disaggregation (tree structure) of the information about the end products to be produced. A product was first classified by type: each type has one or more 'families', each family has one or more 'items'. At each level, a mathematical programming problem is formulated to solve the resulting planning problem. The solution at one level provided constraints to the next lower level. Several authors, including Bitran and Hax (1977) and Axsater (1981), have discussed conditions under which decompositions and aggregations of this kind guarantee that solutions will exist at each level.

Davis and Jones (1988) used this approach to decompose scheduling into a two-level decision-making problem (see Figure 3.6). They use both price-directed and goal-directed methods to ensure coordination of objective functions and constraints. They have also discussed the impact of this on the structure and content of the process plans that are used (Jones *et al.*, 1989) in the scheduling. The top level in this decomposition, the interprocess coordinator, uses coupling constraints to generate limit times – earliest start and latest finish times – for the tasks assigned to the lower levels. Each process module in the lower level solves a sequencing problem using those limit times as additional constraints. Real-time simulation is used to predict the impact of using various scheduling and sequencing rules on each level in the system. An important feature of this work is the possibility of expanding this approach to more than two levels.

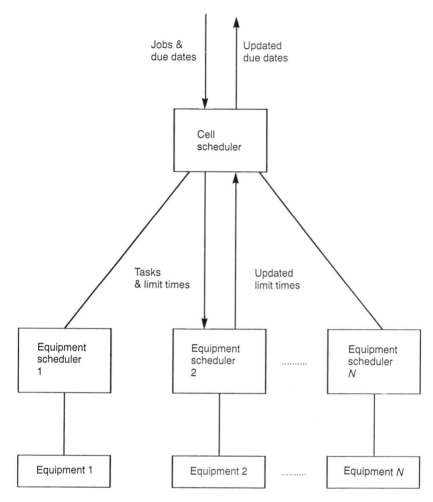

Figure 3.6 *Davis and Jones' two-level scheduler*

4   SHOP-FLOOR CONTROL STRATEGIES

As noted above, cell controllers must fit into some type of shop-floor control system. Two approaches have been proposed: hierarchical and heterarchical.

4.1   Hierarchical Control

Almost all of the proposed shop-floor control architectures use both multi-layer and multi-level control to form hierarchical control systems.

They have a tree structure corresponding to the specific arrangement of equipment in the shop. In addition, the designs seem to be based on three guidelines (Albus et al., 1981): (1) levels are introduced to reduce complexity and limit responsibility, decision making, and authority; (2) each level has a distinct planning horizon which decreases as one goes down the hierarchy; and (3) control resides at the lowest possible level. The application of these guidelines has led to a variety of different architectures. Major differences exist in the number of levels and functions assigned to each level, control paths between supervisors and subordinates, and their handling of data and communications. At the moment, there are no quantitative methods available to compare different designs or to determine the 'best' design for a particular application.

Most implementations of hierarchical control principles have two major limitations. First, all decisions are made at the top level. Controllers in lower levels simply execute one command at a time. Furthermore, these controllers have very limited ability or authority to react to the dynamic evolution of the environment in which they operate. Second, the only exchange of data allowed in this type of system is between a supervisor and its subordinates. This means that a supervisor must transmit all data needed to execute a command along with the command. This lack of peer-to-peer communication means that subordinates are cut off from their chain of command whenever the communication link goes down. In addition, since most controllers can only execute one command at a time, part or all of the system can deadlock very quickly.

4.2   Heterarchical Control

Recently, some researchers (Hatvany, 1985; Dufie and Piper, 1986) have attempted to address these problems. In addition, they seek to eliminate the rigid supervisor/subordinate relationships found in hierarchies. To do this, they have developed theoretical foundations for and advocate the use of heterarchical structures. The principal characteristic of this type of structure is that all entities are treated as cooperating equals. Decisions regarding what to manufacture, how to manufacture, and when to manufacture are made by committee. The committee carries out an extensive and complicated negotiation process to arrive at those decisions. This approach results in 'arbitrary control paths' which, researchers claim, can overcome the potential for system deadlock that exists in most implementations of hierarchical control. Although researchers have demonstrated this approach in the laboratory, they have not shown it to be practical in a real manufacturing environment.

4.3   The AMRF Approach

As noted above, the AMRF has implemented a multi-level shop-floor control hierarchy (see Figure 3.2) using multi-layer and multi-level control principles. The AMRF has addressed both of the limitations described above, not by abandoning the hierarchical control approach, but by implementing advanced technologies.

As noted above, the cell controller (level 3) in the AMRF control hierarchy is the only controller that makes any real-time decisions – it does scheduling. All other decisions about how and when to perform a particular activity are made offline. Recently, AMRF researchers have begun to integrate distributed decision making into the existing hierarchy. Each controller will eventually choose, from alternatives given in a process plan, *how* to complete assigned jobs. It will then use that plan, together with start and finish times from its supervisor, to determine an exact schedule. The framework outlined in Davis and Jones (1988) will be used to distribute both of these decisions across the AMRF shop-floor hierarchy.

As for the peer-to-peer communication problem, this came about because of hard-wired communication links. Since communication could occur only between two entities that were physically wired together, it was necessary to transmit all data needed to do a job. For example, suppose a cell wanted to send a command to a machining centre to machine a part. The cell had to tell the machine tool controller what to do, when to do it, and how to do it. That is, the control path and the data path were the same physical wire. The AMRF researchers recognized early on that the main difficulty in implementing separate control and data paths was not hierarchical control, it was technology. They recognized that more sophisticated computing and communications technologies would be forthcoming. So they designed separate architectures for data management, shop-floor control, and network communication (Barkmeyer, 1989). This, in effect, allows the control paths to be hierarchical and the data flow paths to be completely arbitrary. This means that (1) information can be exchanged between modules anywhere in the AMRF and (2) control interchanges can be restricted to a supervisor and its subordinates.

5   A NEW CELL CONTROL ARCHITECTURE

In this section, we describe the cell controller as one module inside a hierarchical, shop-floor control architecture. We also include a list of additional assumptions and discuss the influence of the AMRF on this architecture. We conclude with a description of the external interfaces and internal implementation structure for the cell.

5.1   Major Assumptions

Each cell controller must be viewed as one part of a larger shop-floor control architecture. We assume that architecture has a tree structure like the one shown in Figure 3.2. As shown, each module is simultaneously a supervisor to many subordinates and subordinate to one supervisor. Each module tries to make optimal use of subordinate resources to complete jobs assigned by its supervisor. As described in Jones and Saleh (1990), each module will eventually perform three functions:

- planning: generate and update a plan for executing assigned jobs;
- scheduling: evaluate proposed plans, generate/update schedules;
- regulation: interface with subordinates, monitor progress.

These functions generalize those developed in Albus (1981) and Saridis (1985). In most applications, they will be performed at different frequencies – regulation the most frequent and planning the least frequent. Hence, we can treat each function as a separate layer in a multi-layer controller. The frequency with which they are actually executed is implementation dependent.

We assume, for the sake of this discussion, that cell controllers form the second level in the shop-floor control system. We stress, however, that this approach is compatible with any of the definitions given above. Each cell has a fixed set of subordinates, the equipment level controllers. We assume that both the cell and equipment controllers execute the three functions described above. If equipment controllers cannot perform these functions (which is the case in many systems today) then the cell must take on this responsibility as well. The cell's supervisor can also be a multi-layer controller, but this is not necessary. We need only assume that the cell receives a list of jobs to do with due dates and priorities.

5.2   Cell Functions

5.2.1   Planning. The planning function (PF) determines a new 'production plan' for each job assigned by the cell supervisor and updates an existing production plan to account for unexpected problems with subordinate equipment (see Figure 3.7). A production plan contains:

- a list of tasks which must be executed to complete the assigned job;
- task assignments for each piece of equipment;
- any precedence constraints among the tasks;
- proposed start and finish times for each task;

The tasks in a given production plan become the jobs for the equipment controllers. They in turn will plan, schedule, and execute all of the operations necessary to complete each task.

For a new job, this involves several steps. First, the PF retrieves or generates one or more candidate production plans. In the near future these candidates will be contained in the process plan (see below) for the job. Later, the PF may have the intelligence to generate these candidates in real time. These candidates are passed to the scheduling function (SF) for estimates of their impact on the evolution of the system. The estimate is based on one or more performance criteria specified by the PF. The selected

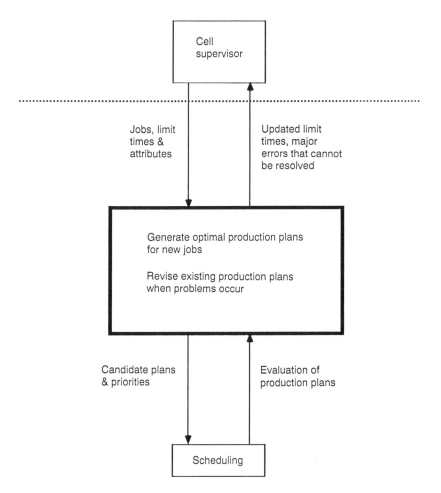

Figure 3.7   *Planning function (PF)*

plan is put into the database for later use by the equipment controllers in constructing their run-time schedules.

Information is provided by the SF on the status of all jobs and all subordinates. Whenever a problem occurs that cannot be resolved by the SF, the PF must determine a new course of action. It may change job priorities, performance measures, and/or existing production plans. Whenever shop-floor conditions are such that the PF can devise no strategy which does not violate one or more due dates, the cell's supervisor must be informed. It will negotiate a new set of due dates or a new set of jobs, or both.

5.2.2   Scheduling function. The scheduling function (SF) performs three major functions (see Figure 3.8): It evaluates proposed production plans from the planning functions; it generates a schedule containing a list of

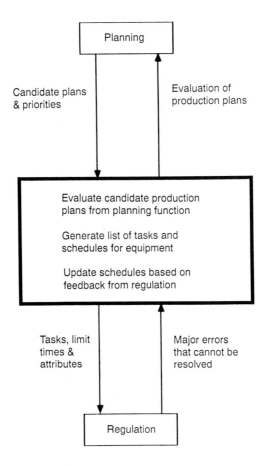

Figure 3.8 *Scheduling function (SF)*

tasks with start and finish times for each equipment controller; and finally, it tries to resolve any conflicts and problems with the current schedule identified by the regulation function (RF).

As discussed above, the SF evaluates candidate production plans for each job. We expect this evaluation to be carried out using a simulation analysis for a specified period into the future. This analysis is performed to determine the impact of a particular plan on the forecasted evolution of the system during that period, that is the schedule. The PF provides the performance measures and candidate plans to be used in the evaluation, and its best guess regarding the up/down time of all equipment during the analysis period. Performance measures can include tardiness of current jobs, utilization and capacity of equipment, load on the system, and throughput, among others. The SF will prioritize candidate plans based on the selected performance measures. Once the PF selects the production plan to be used it must notify the cell's supervisor of the expected completion time for that job.

Before tasks in a particular production plan can be released to equipment controllers, the SF must compute the anticipated start and finish times of those tasks. That is, it must update its current schedule using the performance measures and scheduling rules provided by the PF. These times will be used by the equipment controllers in determining their own plans and schedules and are also passed up to the PF as part of the feedback information.

Manufacturing equipment is subject to random failures which cause delays. These delays, if they are long, can make the current schedule infeasible. The SF must resolve these infeasibilities as quickly as possible. A two-step process is envisioned. First, the number of times in the current schedule which will be impacted by this delay must be determined. The outcome of this analysis determines the second step. In some cases there may be enough slack in the original schedule to absorb the ripple effect of the delay. In other cases, a new schedule can be generated (Davis and Jones, 1988) by simply selecting a new rule from the existing candidate list. Whenever this cannot be done, PF and SF may negotiate new start and finish times, the PF or SF may change the existing performance measures, or the PF must specify a new production plan.

5.2.3   Regulation function. The regulation function (RF) is the interface between the cell controller and its subordinate equipment controllers (see Figure 3.9). It releases jobs to subordinates, monitors subordinate feedback on those jobs, and informs the SF of any problems. The job release strategy depends on the capabilities of the subordinate. If the subordinate can only manage one job at a time, which is the case with most equipment controllers today, then the RF will release one job at a time. A new job is released when the previous one is completed. Feedback data from subordinates is compared with the current schedule to determine if any unexpected conditions have arisen. Information on how this interface actually works is described in later sections.

5.3   Influence of the AMRF

Much of the design described above is based on the cell controllers built in the AMRF. The first cell controller was built in 1983 (Jones and McLean, 1984). Figure 3.10 shows the internal structure and external command/feedback interfaces. At that time there were only three workstations. In addition to these workstations, the cell also interfaced with the data administration system (predecessor of IMDAS) and the network communications system. The cell performed three major functions: QCM (Queue Configuration Manager), scheduling, dispatching. The QCM assigned incoming jobs from the operator terminal to each workstation and

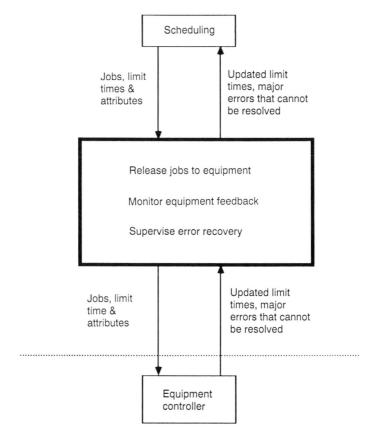

Figure 3.9   *Regulation function (RF)*

provided feedback to the operator on the progress of each job. Since each job could be processed completely at one workstation, the QCM essentially formed three independent queues of tasks. There was one scheduler and one dispatcher for each workstation. Each scheduler sequenced the tasks in its own queue using simple rules such as first in–first out, earliest due date, shortest processing times, etc. Each dispatcher issued the next task in the sequence to its assigned workstation and monitored the feedback from that workstation. Feedback included information on the status of the workstation and the task it was performing. Tasks were dispatched one at a time and only after the preceding one was completed.

Each of the modules in the original cell controller was implemented using state tables. Figure 3.11 shows the simple state table which was used in one of the robot controllers. These provided an easy way to implement deterministic decision logic. Based on the current internal state, supervisory command, and feedback information, the table would specify the next internal state, command to all subordinates, and feedback to the supervisor.

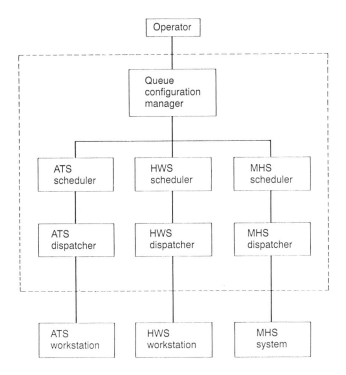

Figure 3.10 *First AMRF cell controller*

The major drawback is that state tables get large and complicated very quickly.

The second version of the cell (McLean, 1987) controlled six workstations (implemented in 1986) with each job going to two or more workstations. Its new design was based on the ideas in Jones and McLean (1985). It differed from the original design in two ways. First, to account for the fact that a job could now require tasks to be done at more than one workstation, the function of the QCM was expanded to (1) retrieve a process plan from IMDAS for each assigned job and (2) parse that plan to determine the ordered sequence of workstations to be used in completing that job. Second, there was only one scheduler for the entire cell. It generated a queue of prioritized tasks which each dispatcher issued (again one at a time) to its assigned workstation. This new version also used state tables, but not to the extent they appeared in the earlier version. In addition, it used spreadsheet technology to display results on a personal computer.

5.4   Impact of Material Handling

Material handling systems (MHS) have a significant impact on the dynamics of a manufacturing cell. They are the primary source of coupling

Command	State	Feedback	Next State	Output	Report
—	C30	No New Command	C30	Wait	—
Fetch (A)	C30	New Command	C31	Reach to (A)	—
"	C31	Distance to A >T1	C31	Reach to (A)	—
"	C31	Distance to A ≤ T1	C32	Grasp (A)	—
"	C31	A Not Visible	C35	Search for (A)	—
"	C32	Grasp Pressure <T2 Grip Dist >T3	C32	Grasp (A)	—
"	C32	Grasp Pressure ≥ T2 Grip Dist > T3	C33	Move to (X)	—
"	C32	Grip Dist ≤ T3	C36	Back Up (Y)	Object Missing
"	C33	Distance to X >O	C33	Move to (X)	—
"	C33	Distance to X = O	C34	Release	—
"	C34	Grip Dist <T4	C34	Release	—
"	C34	Grip Dist ≥ T4	C30	Wait	Report Fetch Done
"	C35	A Not Visible	C35	Search for (A)	—
"	C35	A in Sight	C31	Reach to (A)	—
"	C35	Search Fail	C30	Wait	Report Fetch Fail
"	C36	Back Up Not Done	C36	Back Up (Y)	—
"	C36	Back Up Done	C35	Search for (A)	—

Figure 3.11   *State table to implement the robot command 'Fetch (A)'*

and can propagate delays if they are either overloaded or down due to failures. They can also play a major role in dissipating delays if other equipment is down. Hence, they are a major issue in cell controller design.

In most manufacturing cells, one of two types of MHS are used: discrete (such as a robot or an AGV) and continuous (such as a conveyor) (Tompkins and White, 1984). Continuous MHS are more suitable for serial flow arrangements and are highly inflexible, while discrete MHS are more flexible but are more challenging from a control perspective.

Consequently, the PF and SF must treat these MHS as another finite capacity resource to plan and schedule (Egbelu and Tanchoco, 1984). We believe that, from a cell control perspective, two separate MHS are desirable: one for intracell and one for intercell activities. We point out, however, that the scheduling of MHS in this type of distributed, integrated architecture has not been addressed. To do this, the class of scheduling problems must be expanded to consider MHS as another resource to schedule at all levels. The cell's supervisor schedules intercell material transfers to cell load/unload stations. Each cell controller then schedules internal pickup and delivery times. The transporter scheduler uses these times to decide which transporter to use and the path, if applicable, to be used in completing the transfer. We anticipate that this approach will be tried in the AMRF using the framework described in Davis and Jones (1988) together with the technique illustrated in Erschler *et al*. (1986).

We note that this approach is not used in many existing facilities. Problems arise when the cell tries to plan and schedule around a critical resource that it does not own and cannot control. This can lead to the situation where no feasible schedule can be generated. AMRF researchers have recently suggested that, in situations like this, it may be beneficial to think of the MHS as a 'service' which must be shared by all. This is the same view that led to the development of the IMDAS. Perhaps a separate architecture should be developed for the MHS? Additional research is needed to address this question.

5.5   Impact on Process Plans

Process plans contain the information needed to manufacture, transport, and inspect parts. In the long run, both the cell- and equipment-level controllers will use process plans to determine how to execute assigned jobs. This requires several changes in existing process plans. First, they must have a multi-level structure which parallels the control structure. Second, process plans must provide alternate processing sequences with precedence constraints and allow backtracking when problems occur. This information is needed by both the scheduling and planning functions. Third, plans at different levels must have the same internal structure (AND/OR graphs are one possibility). This simplifies the software development. Finally, since computers will be responsible for processing it, this information must be provided in a consistent, error-free, and machine-readable format.

Recall that a 'job' at the cell level is made up of the tasks to be executed by the various equipment in the cell. The cell-level process plan will contain a 'routing' for each job assigned to the cell. That routing will be a list of the equipment in the cell which can be used to manufacture the part. This includes the precedence relations that determine the alternative sequences in which that equipment can be used. The planning function selects the run-time production plan from those alternatives. The process plan also contains timing information used by the scheduling function. The processing time for the job is the sum of the durations of the tasks that make up that job. Hence the timing data in the cell process plan will be the aggregation of the information in the equipment-level process plans. Equipment-level process plans will contain the programs needed by the equipment to machine a part, move a part, inspect a part, etc. That is, they will contain NC programs, robot programs, inspection programs, etc.

5.6   External Interfaces

The cell control module interfaces with its subordinates and supervisor through some type of command/feedback structure and with the data management system and the communications system.

5.6.1   Command/feedback interfaces.

The cell controller must interface with its equipment controllers and its supervisor. The interface must allow for the assignment, execution, and monitoring of parallel activities. In addition, it must provide for start-up and shutdown of all subordinates. One such command structure contains three top-level fields: ACTION_VERBS, JOBS_POINTER, and RESOURCE_POINTER. Each module will have a valid set of ACTION_VERBS which initiate functions such as STARTUP, SHUTDOWN, and EXECUTE_JOBS. The JOBS_POINTER parameter is a pointer to a list of jobs in the database. Each entry in this list contains a job-type flag (NEW or OLD), a job ID, a job action to be taken (EXECUTE, CANCEL), process plan_IDs, priority, and limit times. The last field in this command structure is the RESOURCE_POINTER which is a reference to another list in the database. Each entry in this list refers to a specific resource request from a subordinate. It also contains the supervisor's response value (ACKNOWLEDGED, ALLOCATED, UNAVAILABLE, COMPLETED, CLEARED, etc.), and the expected time of availability.

A possible feedback structure also contains three top-level fields: OPERATIONAL_STATUS, JOBS_STATUS, and RESOURCE RE-QUEST. The first field, OPERATIONAL_STATUS, indicates the current operational status of the control module. The JOBS_STATUS field is used to report the evolution of all jobs assigned to the module. RESOURCE_REQUEST reports ordinary or emergency run-time resource requests to the supervisor. Each of these three fields will be further divided into two subfields: CONDITION and POINTER. The former consists of a simple set of ASCII responses and the latter is a pointer to a more detailed list in the database. This structure reduces the complexity involved in implementation by fixing the number of input parameters and by limiting the number of values that each of those parameters can take on.

5.6.2   Data management interface.

The cell and its equipment-level subordinates need a wide variety of data to carry out their functions. There are typically three ways to access that data: each cell gets the necessary data from its supervisor and passes it to the equipment controllers; each cell has its own database management system and passes retrieved information to the equipment controllers; each controller must interface with a global database management system. The first alternative is widely used today, particularly by PLC-based cell controllers. The second alternative is frequently used by the cell controllers marketed by 'system integrators'. These, we believe, provide a short-term solution only. The last choice is the only viable one for the future. There are three major characteristics of future manufacturing systems which support this conclusion.

First, the manufacturing environment is likely to be a heterogeneous one with equipment and cell computers purchased from a variety of vendors. This means that it is necessary to provide users with a common method of accessing data, and to perform whatever translation, assembly, and conversion is needed to fill user requests. Second, there will be some (possibly a large number of) parts which have operations performed in more than one cell. The data needed to make these parts must be shared across those cells. This means that the data system must both enable asynchronous interchanges of information between cell computers anywhere in the system and allow for the replication of some information units on two or more systems and the frequent and timely updates of those units. Finally, data delivery, like material delivery, is not instantaneous. It must be included in the planning of each production job. This means that data is quickly becoming a critical resource which must be scheduled. Furthermore, the scheduling decisions made by the data manager must be coordinated with scheduling decisions made at the planning and scheduling layers of the cell.

This 'separate data system' approach has an added advantage because it allows research on the two systems – control and data – to proceed independently, provided their interrelationships are well understood and accounted for in the final designs.

5.6.3   Data communications. Communication requirements for the components of each cell are typically of two types: very frequent command and feedback messages, and less frequent interchanges with the data management system. The command feedback messages contain relatively small amounts of data, while data interchanges contain relatively large amounts of data, such as process plans or NC programs. We believe that the physical architecture best able to handle this situation is a backbone network integrated with a subnet for each cell. The backbone network is used for intercell communications, access to each cell's supervisor, and the data management system. Each subnet is used for intracell communications. This requires each physical subnet to be transparently connected (using routers, bridges, and gateways) so that any process could conceivably communicate with any other process anywhere in the system. In addition, we recommend that messaging and protocol standards, where available, be used in the design and implementation of every component in the system.

We note three advantages to this approach. First, subnets make it easier to meet the timing requirements for tightly coupled, intracell command/feedback transfers. Second, one can select the physical medium, access mechanisms, topology, and protocols for each subnet to be tailored to the needs of the particular cell it serves. Third, the backbone network is not cluttered with intracell communications which can negatively affect the response time of processes accessing the data management system.

5.7   Internal Cell Implementation Structure

Figure 3.12 shows one possible internal implementation structure for the cell controller. It is based on the AMRF work described in McLean

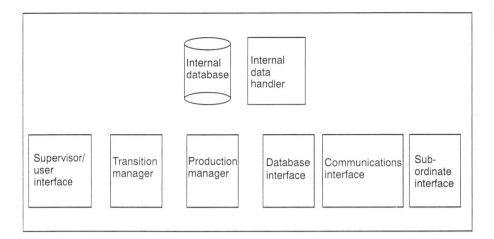

Figure 3.12   *Internal cell structure*

(1987) and Catron and Thomas (1988) and is best implemented on a system with a multi-tasking operating system.

The supervisor/user interface module contains the software needed to interface with the supervisor. The supervisor can be either a supervisory controller or a human operator. This module retrieves input commands from the communications manager, parses them, and passes the appropriate information to the planning module in the production manager. It also builds the feedback messages from the information provided by that planning module and forwards those messages to the communications manager to be sent back to the supervisor.

The subordinate interface module performs similar functions. It takes information from the production manager and builds the commands to be issued to the subordinates. It also parses the feedback from the subordinates and provides that feedback to the regulation module in the production manager. Depending on the application, there may be only one subordinate interface which handles the command/feedback messages for all subordinates or separate modules for each subordinate.

The transition manager module includes the software needed for initialization, start-up, error recovery, and shutdown. This provides the internal synchronization needed to start up and shut down both the cell itself and all its subordinates.

The production manager module contains the software needed to carry out the functions described in the planning, scheduling, and regulation layers of the cell controller. The decisions at both the planning and scheduling layers are stochastic in nature. This happens because uncertainties arise from the aggregation of information and increase in planning horizon that takes place as one moves from the equipment level to the cell level. The decisions at the regulation layer, on the other hand, are essentially deterministic. This happens because the RF assumes that subordinates will execute assigned tasks according to the prescribed plan and current schedule. Mathematical programming, simulation and expert systems have all been proposed and used to carry out these functions. We note that the framework described in Davis and Jones (1988) is attractive because: it combines the best features of all of these techniques; it already includes negotiation and compromise analysis; and it can be used at all levels in the proposed architecture.

All requests to retrieve and/or update process plans, schedules, and other data that reside in the global database must go through the database interface module. That module poses the necessary 'queries' in the format expected by the data administration system. When the response comes back, this module parses the incoming message and informs the internal data handler that new data has arrived.

The data handler translates all incoming data into the internal formats needed by the production manager module and stores them in the internal database. It also revises that database as needed and performs the operations necessary to convert that data back into the form used by the database to execute its updates, consistency checks, etc.

The communications interface is responsible for sending and receiving all command/feedback messages between the cell and its subordinates and all interactions with the global database. This module is the interface with the network communication system and must implement all required protocols. It must initiate the communications when the cell controller 'comes up', and terminate communications before the cell 'goes down'.

6 SUMMARY

In this chapter, we specified an architecture for a cell controller that provides a high degree of intelligence and can be easily integrated into most hierarchical shop-floor control systems. It performs planning, scheduling, and regulation. We also examined cell information, data management, and communication requirements.

REFERENCES

Albus J. (1981). *Brains, Behavior, & Robotics*. New York: McGraw-Hill.
Albus J. S., Barbera A. J. and Nagel R. N. (1981). Theory and practice of hierarchical control. *Proc. 23rd IEEE Computer Society International Conference, September*.
Axsater S. (1981). Aggregation of product data for hierarchical production planning. *Oper. Res.*, **29**, no 4, pp. 744–756.

Barkmeyer E. (1989). Some interactions of information and control in integrated automation systems. *Advanced Information Technologies for Industrial Materials Flow*. New York: Springer, pp. 39–53.

Benders J. F. (1960). Partitioning procedures for solving mixed-variables programming problems. *Num. Math.* **4**, 238–52.

Bitran G. R. and Hax A. C. (1977). On the design of hierarchical production planning systems. *Decis. Sci.*, **8**, 28–54.

Catron B. and Thomas B. (1988). Generic manufacturing controllers. *Proc. IEEE Conf. on Intelligent Control*, Arlington, Virginia, August.

Dantzig G. B. and Wolfe P. (1960). Decomposition principles for linear programs. *Manage. Sci.*, **8**, 101–11.

Davis W. and Jones A. (1988). A real-time production scheduler for a stochastic manufacturing environment. *Int. J. CIM*, **1**, 101–12.

Duffie N. and Piper R. (1986). Non-hierarchical control of a flexible manufacturing cell. *Proc. Int. Conf. on Intelligent Manufacturing Systems*, Budapest.

Egbelu P. J. and Tanchoco J. M. A. (1984). Characterization of automated guided vehicles dispatching rules. *Int. J. Prod. Res.*, **22**, no 3, p. 359 (26).

Erschler J., Roubellat F. and Thomas V. (1986). *Real-Time Production Scheduling for Parts with Limit Times*. Technical Report No 86063, Laboratory for Analysis of Automated Systems, Toulouse.

Geoffrion A. (1970). Elements of large-scale mathematical programming, parts I and II. *Manage. Sci.*, **16**, 652–91.

Gershwin S. B. (1989). Hierarchical flow control: a framework for scheduling and planning discrete events in manufacturing systems. *IEEE Proc. Special Issue on Discrete Event Systems*, to appear.

Hatvany J. (1985). Intelligence and cooperation in heterarchical manufacturing systems. *Rob. CIM*, **2**, 101–4.

Hax A. C. and Meal A. C. (1975). Hierarchical integration of production planning and scheduling. *Studies in Management Science: vol. 1, Logistics* (M. A. Geisler, ed.), North-Holland, New York.

Jamshidi M. (1983). *Large-Scale Systems: Modeling and Control*. North-Holland, New York.

Jones A. and McLean C. (1984). A cell control system for the AMRF. *Proc. Int. ASME Conf. on Computers in Engineering*, Las Vegas, Nevada, August.

Jones A. and McLean C. (1985) A production control model for the AMRF. *Proc. Int. ASME Conf. on Computers in Engineering*, Boston, MA, August.

Jones A. and McLean C. (1986). A proposed hierarchical control model for automated manufacturing systems. *J. Manuf. Sys.*, **5**, 15–25.

Jones A. and Saleh A. (1990). A multi-level/multi-layer architecture for intelligent shop floor control. *Int. J. CIM*, **3**, No 1, 60–70, special issue on intelligent control.

Jones A., Barkmeyer E. and Davis W. (1989). Issues in the design and implementation of a system architecture for computer integrated manufacturing. *Int. J. CIM*, special issue on CIM architecture, **2**(2), 65–76.

Libes D. and Barkmeyer E. (1988). The integrated manufacturing data administration system (IMDAS) – an overview. *Int. J. CIM*, **1**, No 1, 44–49.

McLean C. R. (1987). A cell control architecture for flexible manufacturing. *Proc. of the Advanced Manufacturing Systems Conf.*, Chicago, Illinois, June.

Mesarovic M. D., Macleo D. and Takahara Y. (1970). *Theory of Multilevel Hierarchical Systems*. New York: Academic Press.

Rybczynski S. *et al.* (1988). *AMRF Network Communications.* NBSIR 88–3816, National Institute of Standards and Technology, Gaithersburg, MD, USA.

Sandell N., Variya R., Athans M. and Safanov M. (1978). A survey of decentralized control methods for large scale systems. *IEEE Trans. Autom. Control*, **AC-23**, no 2, 108–128.

Saridis G. (1985). Foundations of the theory of intelligent controls. *IEEE Workshop on Intelligent Control*, Troy, New York, August.

Simpson J., Hocken R. and Albus J. (1982). The Automated Manufacturing Research Facility of the National Bureau of Standards. *J. Manuf. Syst.*, **1**, no 1, 17–32.

Tompkins J. A. and White J. A. (1984). *Facilities Planning.* New York: Wiley.

Villa A. and Rossetta S. (1986). Towards a hierarchical structure for production planning and control in flexible manufacturing systems. *Modeling and Design of Flexible Manufacturing Systems* (Kusiak A., ed.). Amsterdam: Elsevier.

4 *The cell management language*

David Alan Bourne

The cell management language (CML) is a powerful computing environment for building complex manufacturing systems. Furthermore, CML does not assume that system components are compatible, because the best system components usually come from a mix of different vendors. And these same vendors often explicitly design to prevent multiple vendor systems.

CML combines computational tools from rule-based data systems, object-oriented languages and new tools that facilitate language processing. These language tools, combined with rule processing, make it convenient to build new interpreters for interfacing and understanding a range of computer and natural languages. This ability allows CML control programs automatically to understand and generate part programs in different languages.

Manufacturing systems require many cooperating subsystems in order to plan and control a wide range of manufacturing tasks. To facilitate this goal, CML makes it easy to link together interpreters for each subsystem into a single control system. As one control system, CML can manage and control system-wide constraints and interactions.

CML has been used in the factory environment since 1984 to make preform turbine blades and has proven to simplify greatly the task of building new manufacturing control systems.

1 INTRODUCTION

The development of advanced robotics brought expectations of increased productivity and quality control, but to everyone's disappointment, these expectations still have not been realized. Advanced standalone machines have not greatly improved productivity, and integrating large systems has been prohibitively expensive. What is worse, the few integration projects that have been undertaken took inordinate amounts of engineering time; there have been several projects that have taken more than 50 man years to complete. Most of this time has been spent by engineers trying to put round plugs into square sockets. Some machines have not been designed to allow for any communication, while others provide only partial communications that presume a person is operating the front panel. The few machines that are supplied with communication systems are rarely compatible even within a single vendor. However, the buyer should be aware that this is rarely apparent from trade show demonstrations, advertising brochures or sales people.

At least one dream for factory automation should be simple: roll a computer on to a factory floor, plug it into a set of machines from different manufacturers, start a program, and then, with absolutely *no* traditional programming, begin the task of integrating the machines into a cooperative cell. The cell management language (CML) brings us within reach of this dream. Its pursuit demands that CML copes with a range of approaches that are encountered in manufacturing.

(1) *Generic vs custom.* The software tools for building flexible manufacturing systems must be generic, so that the same tools can be used over and over again to solve new system integration problems. Also, these tools must be powerful enough to reduce drastically the development time of the overall system. As a rule of thumb, our laboratory has set a goal of increasing productivity for building software for automation by a factor of thirty* over existing approaches to customizing new applications.

(2) *Mixed equipment vs standardized.* What makes our approach different from most others is that we are building systems that work with existing equipment, and make only a new minimal assumptions (e.g. nominal communication abilities (Fussell *et al.*, 1984)). This contrasts sharply with a scheme that forces every piece of equipment to fit within a standardized framework.

(3) *Intelligent vs hard-coded.* The final system should be autonomous and should be able to make its own decisions about the actions it takes in both normal and abnormal conditions (Bourne and Fox, 1984). For example, it is not sufficient to have 'hard-coded' responses to fixed situations, because there will always be some new unanticipated event and situation. A vision system for locating parts is an architectural solution for removing 'hard constants', and inference rules that can be used to deduce the reason for a machine's failure are a software solution to removing 'hard procedures'.

In addition to satisfying these more abstract goals, CML (the control language) must provide an environment in which programming manufacturing systems is drastically easier than it is today. To accomplish this, CML provides tools for automatic programming in different target languages (e.g. a vision machine language and a robot language) and facilitates the teaching of rules for controlling multiple machine interactions (see also

---

* The magic number of thirty recognizes that it takes about 30 days to implement an idea that it takes 1 day to think of in the first place. The ideal situation would be to eliminate essentially all of the programming time.

<Tablename>

<Tablename>	<Fieldname-1>	...	<Fieldname-N>
<Entryname-1>	<Dataitem-1-1 or Instruction-1>	...	<Dataitem-1-N>
...	...	...	...
<Entryname-M>	<Dataitem-M-1 or Instruction-M>	...	<Dataitem-M-N>

Figure 4.1 *The table and its parts*

Wright and Bourne, 1988). These tools include a parser driven by a grammatical description (syntax), rule primitives (semantics) and database manipulation primitives (pragmatics). These are the most basic elements for defining, interpreting and generating text in different languages.

CML is a table-oriented environment that represents both programs and manufacturing data. This has proven to be a natural way to represent most of the information found in manufacturing. In addition, instructions that manipulate tables are easily 'pictured' by people not accustomed to abstractions forced by other languages. For example, it is possible to update an entire table with data from another table, or it is possible to treat the values in a table as numbers that can be added up down the columns like a spreadsheet.

The two-dimensional tables in CML can be referenced and manipulated as a whole or in part (i.e. entries, fields and items). The structure of a table and the names of its parts are shown in Figure 4.1. One benefit of this uniform representation is that both programs and data can be updated with database operations.

Tables are a convenient representation for both database and language processing tasks, which are common in manufacturing. For example, the input and output in a complex manufacturing system has to be carefully crafted to suit a target machine or application. In this case, a grammar that describes the language of the target (e.g. a machine tool) can be embedded in a table and then used to drive a general-purpose language generator or parser. In fact, both text generation and parsing are standard CML operations.

To accomplish database operations, CML operators can be used to access and modify tables directly. However, if the user has a strong preference for a particular data model then higher-level functions can be constructed to emulate the behaviour of a relational database or a more complicated inheritance structure, which is typically found in an expert system shell.

To build a CML application, it is necessary to define a set of tables that define the various parts of a program and application data. These tables are always defined as being part of a *workspace*, a named collection of tables, so that separate applications can be maintained to provide an additional level of data abstraction. It is also convenient to think of CML workspaces as objects in an object-oriented environment, because it is possible to manipulate them and communicate with them as an object. From an application perspective, it is customary for each machine in a manufacturing system to have its own CML workspace and so the workspace takes on the identity of the real-life machine. We will see the relative role of tables and workspaces better as the chapter develops.

2 BACKGROUND

CML is designed to integrate systems by using software to overcome the many incompatibilities that are found in the factory environment. Other research groups use different methods to integrate factory systems. The principal leaders in this area are General Motors Corporation and the National Bureau of Standards (now NIST).

2.1 General Motors and MAP

The Manufacturing Automation Protocol (MAP) is a multi-vendor initiative to standardize the communication protocols between the computational components in the manufacturing environment (Veilleux and Petro, 1988). The overall scope of the project is to standardize the interconnections of equipment from the level of the communication medium all the way up to a complex message-passing scheme. The goal is to allow for file transfers and other application-dependent tasks within a multi-vendor network. The automation of industry today depends on this kind of standardization effort.

CML enters the manufacturing arena with solutions beyond the scope of the MAP initiative. The MAP protocol offers a standardized approach for getting digital information from one place to another, while CML offers the intelligent understanding and generation of these messages. The problem is that machine tools, robots, vision systems and people all use different languages to express their ideas and actions, and understanding the content of a message often involves making sense of a computer program. CML allows users to build easily a number of program-understanders and program-generators that can be integrated into a system to control the whole cell.

2.2 The US National Bureau of Standards and AMRF

The Advanced Manufacturing Research Facility (AMRF) project also has a goal of standardizing multi-vendor communications (Simpson *et al.*, 1984). This proposed communication model is similar to, but not identical to, MAP. The AMRF communications proposal has not been finalized, but it is only a matter of time before there is a consensus between the MAP and AMRF proposals.

NIST also is developing a general architecture for factory automation. Briefly, it proposes a hierarchical control system with finite state machines representing the actions of each node in the hierarchy. In addition, there is a system-wide clock that causes the nodes of the hierarchy to update their state on a regular basis. Even through NIST has developed a strategy for control problems in manufacturing, it has not developed a wide base of generic computational tools.

While CML is a general-purpose language that can be used to implement the NIST control strategy, we have found more effective solutions. Principally, CML is built to perform transaction-based processing rather than polling for state changes. This allows the system to be more responsive to demanding situations, because the CPU cycles are not wasted on polling and are only used on an as-needed basis. In any event, the actual control of manufacturing activities is only a small fraction of the responsibilities taken on within a CML environment. The primary emphases of CML solutions involve the automatic planning and generation of part programs.

2.3   History behind CML

In 1981, Carnegie Mellon University and Westinghouse Electric Corporation started working on a large project to automate the production of preformance turbine blades for steam generators (Wright et al., 1982). This system integration problem presented several challenges unlike those encountered in most commercial flexible manufacturing systems at the time. In such a system, there is a strong assumption that machines are run independently of each other (Wright and Bourne, 1988).

The machines for the open-die forging process, also known as swaging, were manufactured by companies that are not involved in the modernization of controls and had to be completely reworked: an open-die forge, a large rotary furnace and two heavy-duty robots. Therefore, most of the effort in the first phase of the project went into the development of the controls at the machine level. In one case, we had to rewrite the controls from scratch. By the time we were ready to integrate the system, most of the allotted project time (through March 1983) had been exhausted. Making matters worse, a 16 bit computer had been selected as the cell host. As a result, a disproportionate amount of our time and effort had to be spent solving memory management problems. Under these circumstances, we were only able to make the cell host into a general node for message passing with rudimentary logical control. The messages that were passed in the system were hard coded, and the cell host had no real understanding of their content.

Hot-metal working is a hazardous application and so it is frightening to think that the control system did not understand the messages that it was sending. At first, we were forced to run the machines in strict sequence so that a human could intervene if necessary. This precaution was a direct result of the host not having enough intelligence to understand what was happening in an unforeseen situation. By 1983, everyone involved in the project had agreed that a new approach to the project would be required to satisfy our goal of automating an unmanned cell in a dangerous working environment.

Work began in April 1983 to design a programming language that could both understand multiple languages and communication protocols, and provide a simple and extensible approach to system coordination. This language became CML.

In the autumn of 1984, our partially automated manufacturing cell at Westinghouse was converted to run as a CML application. However, by this time, some of the equipment in the cell was also being upgraded, which could have caused an enormous amount of reprogramming if we had pursued a traditional strategy. During the winter of 1985, the two robots in the cell were replaced with two new robots and two new controllers. The new software was written and tested before the robots were bolted to the floor. In 1986, we added two more machines to the cell, and only a handful of additions to the CML control program were necessary. To use these two new machines in the cell effectively, we had to add a few entries to system tables and then write a few rules for how they would interact with the other system components. By far, most of the time was spent at the machines themselves to understand their functionality.

The result of this effort was a general set of tools that can be used to build a 'universal interface' in the manufacturing environment. Figure 4.2 shows how CML can be positioned in the factory environment to solve many system integration problems. At the bottom of the figure is an early picture of the forging cell (*circa* 1982). This cell has since grown to include nine machines that work simultaneously whenever possible.

This cell control system in CML can perform the following manufacturing system functions.

(1) Dynamically upload and download part programs.
(2) Automatically build vision programs based on part definitions.
(3) Dynamically sequence machine operations avoiding deadlocks and collisions.
(4) Operator interface and support functions.

In 1985 and 1986, Westinghouse took CML on as a commercial venture. In this new context, CML was used to build two more large-scale flexible manufacturing systems. Unfortunately in 1987, Westinghouse chose to terminate its automation division, but the implementations using CML in the factory have remained operational.

In the course of the events, computing power (i.e. speed and size vs dollars) has radically increased. It is now planned to revive CML by porting it to a large personal computer platform (e.g. Motorola 68030 based with 8 megabytes and up). The tables in this chapter have been created in this environment. On this platform, we can effectively change the dream presented earlier of 'rolling a computer in and . . .' to 'carrying a computer in and . . .' which in practice makes a big difference.

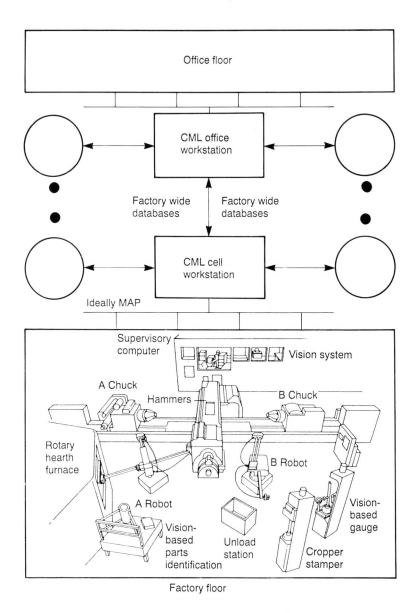

Figure 4.2   *CML's role in the manufacturing environment*

Grammar

Grammar	f1	f2	f3	f4
Grammar	f1	f2	f3	f4
Sentence	Grammar:Question	Grammar:Declare		
Question	verb/label="-quest"	art	noun/label="-subj"	adj
Declare	art/label="-declare"	noun/label="-subj"	verb	adj

Figure 4.3 *Grammar for simple declarative and interrogative sentences*

3   MACHINE SPECIALISTS

The first step to building a control system for a multi-vendor manufacturing cell involves building interpreters, or machine specialists, for each device. This requirement is just one of the reasons why it is drastically more difficult to integrate a multi-vendor system than it is to integrate a single-vendor system. CML provides standard tools for developing these machine specialists, whose job it is to interpret programs and command messages, and to generate programs and command messages. Normally, an interpreter is represented by the following relation:

$$\alpha \rightarrow \beta$$

A program in the language α (e.g. CML) can be translated into a set of possible actions β (e.g. the result of executing a CML program).

CML further augments this idea by allowing the user to set up a number of interpreters within the CML environment. One of these interpreters defines the language α' used by an external device.

$$\alpha \rightarrow \{\alpha' \rightarrow \beta'\}$$

To define a new interpreter, the grammar of this language must be specified, as well as a function that selectively maps these programs into actions β' (e.g. database operations). Within CML there is one command that binds the grammar and the semantic function into one unit. This command loops over each input sentence, applies the grammar that results in a parse table and then applies the semantic function to the parser's output resulting in a final action.

The grammar of α' is defined in a table that is in turn input for the parser. Figure 4.3 shows a grammar that defines simple declarative and interrogative English sentences. The parser starts on the first entry of the grammar (i.e. **Sentence**) and works its way across the items. The first item in that row directs the parser to the row labelled **Question**, where it attempts to find a series of words, separated by spaces, that fit within the appropriate lexical categories defined in Figure 4.4. If this path in the parsing succeeds, then the parser returns to the top entry and usually skips over all of the remaining optional items; this assumes that there is no more input to parse. If the parse

Verb

Verb	number
is	single
are	plural
were	plural
was	single

Noun

Noun	number
parts	plural
part	single
robot	single
robots	plural

Adj

Adj
ready
baking
titanium
steel

Art

Art
the
a
all
each

Figure 4.4 *Lexical tables named in the grammar description*

fails, then parsing is directed to the entry named **Declare** and the same procedure is followed.

The grammar refers to lexical tables (Figure 4.4) that define the legal words for each syntactic slot. This example does not check for grammatical attributes, such as matching a verb's number with its corresponding noun's number. To perform these checks it is necessary either to multiply the number of lexical categories, or to provide an additional level of semantic attachment, or in this case, checking. We have preferred to use rather weak grammars with additional checking being performed by a set of rules. This approach has the advantage of accepting and making sense out of a wide range of sentences that would otherwise be deemed syntactically invalid.

Each time the parser is given input, it uses the grammar description to help it produce an output table. This output table deviates from most other parser's output, because it is a linear representation of the parse tree; these table items are essentially the names of the leaves in a parse tree. In order to keep these leaves unambiguously ordered, special labels (in quotes in Figure 4.3) are attached to the node names to identify the location in the tree. Notice that the entry names in Figure 4.5 are composites of the lexical type and parse tree's label.

Once the input has been satisfactorily parsed (Figure 4.5), a function is automatically executed to attach meaning to it. Typically, this function conditionally fires a set of rules based on the input. For example, in Figure 4.6 there are two rules (or function calls) that fire only in the case when there is a value in the parse output for each argument in the function call. In this case, the second rule **R2** will fire, because there are corresponding entries in the **$Parse** table: **art-declare**, **noun-subj** and **adj**.

Input⟩ *is the part titanium*
Answer⟩ *Yes*
Input⟩ *the part is ready*

$Parse

$Parse	Value
art-declare	the
noun-subj	part
verb	is
adj	ready

Figure 4.5 *Parser input and its output*

Rules

Rules	Eval	Func	Arg1	Val1	Arg2	Val2	Arg3	Val3
R1	eval	lookup	verb-quest	x	noun-subj	y	adj	z
R2	eval	update	art-declare	x	noun-subj	y	adj	z

Figure 4.6 *Simple rule set for semantic attachment*

$Sub

$Sub	Value
art-declare	the
noun-subj	part
adj	ready

Figure 4.7 *Substitution table produced by rule execution*

After it has been determined that the function **update** in the second rule should be executed, a substitution table (Figure 4.7) is built that binds the formal types to their corresponding values. In this way, the function's arguments are also in table format and can be manipulated using the common database mechanisms.

At this point, control would be passed to the function **Update**. This function will replace the value **baking** in the database table **Parts** by the part's new status **ready**, which is found in the substitution table (Figure 4.8). This is accomplished when the command **update** evaluates the address of its first argument (tablename: entryname: fieldname) to a value, which is then used to update the value at the second address. Commands such as this can also be used to operate on whole tables, entries, fields, as well as

Update

Update	Command	Arg1	Arg2
e1	update	$sub:adj:value	parts:order1:status

Parts

Parts	Status	Material	Batch
Order1	baking	titanium	101
Order2	ready	steel	93

Figure 4.8 *Update function and simple database for part descriptions*

supporting alternate addressing schemes that involve indirection and numeric indices.

In this example, we have shown how an input language can be described, parsed, and used to drive rules, which then caused the execution of a database update. Because of the common structure of data in CML, all of the illustrated operations can be used in any step of the process. This allows for possibilities that would be essentially impossible to program in most other languages. For example, a set of rules (like Figure 4.6) could be matched against the **$Sub** table (the function **update**'s parameters). This would make it very easy for a function to determine which parameters had been supplied in the function call and the function could then proceed to track down or solicit missing required data.

Most machine specialists and their parsers are not designed for input produced by a person, but rather are designed to handle intermachine communications. This means that the CML parser must offer a rather unusual set of options to cope with the different technologies offered by multiple vendors. These options include scanning input in different radices, representations, byte orders and other configurations of data.

We have used this tool to understand programs that control robots, vision systems, machine tools and people. We have also found that it is a useful approach to building new tools for enhancing the CML environment. For example, the CML assembler is a CML program that converts a text representation of CML into its internal representation. This would have been a major project in other languages, but was a one day project in CML.

4   PROGRAM GENERATION

An interpreter is usually expected to 'understand' programs and then to take some appropriate action based on that understanding. However, in a limited way, the interpreters described in the last section must also be able to 'speak' the language. This usually involves three additions to an interpreter, making it a complete machine specialist.

- *Specification language.* This language is the internal represent-ation for specifying an external machine's methods, capabilities and goals.
- *Algorithm.* An algorithm that converts the problem specification into the appropriate program structure.
- *Grammar.* A surface description of the grammar that when applied to the program structure effectively converts it into the appropriate surface representation. Finally, this can be sent to the external device and executed.

Conveniently, these three additions are the same mechanisms that are used for program understanding, but are now used to generate programs instead. However, it is usually difficult to make the input and output language mechanisms identical.

CML facilitates automatic program generation in several key ways. Since all programs and specifications are represented in the database, new programs can be built using intertable operations and modified using intratable operations. The intertable operations provide a convenient abstraction of an otherwise detailed program, so that a simple formula can be used to describe how the program segments should be pieced together. An example of such a formula is:

$$program:: = header + repeating\ body + trailer$$

This formula, together with a part description, can be used to generate an algorithm either to machine or to inspect a part.

In this case, the program is accumulated by pasting the *header* on to the table that represents the final program. Then, given a skeletal version of the *repeating body*, the intratable operations can be used to update the key values. These updated program segments can be accumulated on the final program within a control loop. Finally, the *trailer* can be appended, completing the program. The constructive approach illustrated by this formula has been successfully used to generate part programs for machine centres and inspection devices. As a result, these programs no longer have to be written by hand for every part style.

Once a program has been built, it usually has to be transformed into a final surface structure that is acceptable to the external machine. For example, the NC program shown in Figure 4.9 has to be compacted, blocked into fixed block size and preceded by a 'partial program command' (notated 'C,TPP').

This final 'grammatical' transformation is eventually accomplished by two commands. The first step compacts the entire table (Figure 4.9) into one string, and the second command blocks the strings and attaches the required

Program

Program	Block	A1	A2	A3	A4	
e1	N1	G00	G90	B-31.		*Header*
e2	N2	Q03				*Body-1*
e3	N3	G90	X2.75	Y23.	B-45.	
e4	N4	Q03				*Body-2*
e5	N5	G90	X2.75	Y27.	B-45.	
e6	N6	M2				*Trailer*
e7	N7	G00	G90	Y30.	B0	
e8	N8	M30				

Figure 4.9   *Final program structure*

$Parse2

$Parse2	f2	f1
e1	C,TPP	$N1G00G90B-31.*N2Q03*
e2	C,TPP	N3G90X2.75Y23.B-45.*N
e3	C,TPP	4Q03*N5G90X2.75Y27.B-
e4	C,TPP	45.*N6M2*N7G00G90Y30.
e5	C,TPP	B0*N8M30M40

Figure 4.10   *Final surface structure*

syntactic sugar. The result is shown in Figure 4.10, and this is sent entry by entry to the receiving machine controller.

Now that we have developed a machine specialist that can both understand and generate programs, it is possible to intertranslate between multiple languages. This is an important feature in manufacturing, because even when machines use the same control language, the machines themselves are usually different enough to require minor translations. For example, two virtually identical machine tools may have their axles labelled differently, and cutter offsets almost always vary from machine to machine. This is the same problem that two English speakers can have. For example, generational and geographical differences change the language enough to require minor translations.

5   SYSTEM OF INTERPRETERS

Once machine specialists have been built for each machine, their performances must be orchestrated into a coherent system. This system must be responsive to the external environment, and it must be cautious about its choice of action.

Each machine specialist is responsible for synthesising complex messages into atomic units (state names). These atomic units then become the lexical names for a new language: one for the whole cell. In this new language, a process plan can be written for each machine in the cell. In order to accomplish this, however, each logical state must be decomposed into three substates:

- **state-R** – the machine is 'ready' to execute an action;
- **state-W** – the machine has been asked to execute an action and we are 'waiting' for it to complete;
- **state-C** – the machine has told us that it has 'completed' the execution of an action.

Each machine has its own process plan, and as this process plan is executed the current state of each machine is recorded in a global table **Cell-States** (see Figure 4.11).

The messages from a single machine are parsed, and a set of rules are expanded (matched and executed) to assign and execute the message's meaning. The entry names of each table are the general categories: syntactic categories and machine-state categories. The legal values of these categories can be any value that properly belongs as its member. Again, the similar structure of the **Cell-States** table, the **$Parse** table and the **$Sub** table make it very easy to reuse code segments (see Figure 4.12) or to match rule sets.

The **expand** command, in Figure 4.13, associates a data set with a rule set and then 'fires' the appropriate rules.

Figure 4.14 illustrates the internal arrangement of the software making up the system of interpreters. In a typical CML system there are several active processes – CML itself, and processes that manage data from external lines at an interrupt level. In addition to these processes, others can be added that perform application-dependent functions. For example, we have implemented an emulation package that allows us to replace any number of real machines with a program that gives an expected machine tool response.

Robot-Process-Plan

Robot-Process-Plan
Load–Furnace–R
Load–Furnace–W
Load–Furnace–C
Load–Tool–R
Load–Tool–W
Load–Tool–C

Cell-States

Cell–States	Value
Robot	Load–Furnace–R
Furnace	Open–Door–C
Tool	Machining–W

Figure 4.11 *Process plan and cell-state summary*

$Parse3

$Parse3	Value
Art-1	The
Noun-1	catalog
Verb	has
Adj-2	four
Noun-2	programs

Cell-States2

Cell-States2	Value
Robot	Loaded-Part-C
Furnace	Door-Open-C
Tool	Machining-W
Vision	Finding-Part-R

Figure 4.12   *Comparison between message parse table and cell states*

MAIN⟩*expand,Cell-States2,Rules2*

Rules2

Rules2	Eval	Func	Arg1	Val1	Arg2	Val2
R1	eval	act	Furnace	=Door-Open-C	Robot	=Loaded-Part-C
R2	eval	act	Robot	=Loaded-Part-W	Furnace	=Door-Open-C

Figure 4.13   *Rules of interaction*

With this package, the control system for a manufacturing cell can be effectively tested without using real machines.

Figure 4.14 schematically shows interprocess communication via mailboxes. These mailboxes are located in shared memory and are supported by background procedures that provide features similar to those expected of a human-oriented mail system (e.g. sending a message to a named process, forwarding, carbon copies, and logging).

A message (M-1) enters the system at interrupt level and is placed directly in a CML device mailbox. After the message is received, it is automatically transferred (becoming M-2) to the protocol's mailbox, which is sorted by priority. In this way, the procedure can read all of its mail from one point in the program, which avoids having the program wait for mail in one place and then receive it in another. The protocol handler strips off the protocol information and decides whether or not the message is valid. If it is not, it informs the sender. Otherwise, it forwards the data portion (M-3) of the packet to the CML agenda mailbox.

At this point, all of the data processing is done within a CML application program. The application program reads the next task on its agenda, and determines how it should be processed by looking at routing information in the **System Dispatch** workspace. In the dispatch workspace there is a table called **$Lang** (Figure 4.15) that provides appropriate information, including the name of the workspace that specializes in this kind of message, the name of the grammar table to parse it and the function to be evaluated after

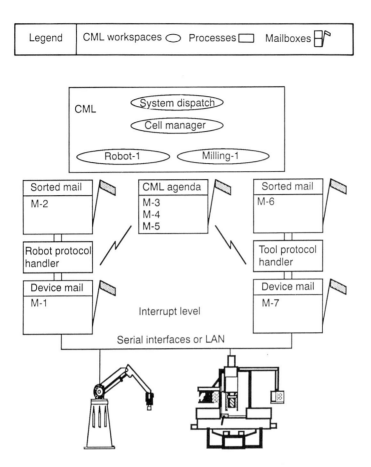

Figure 4.14 *Data flow in a CML system with robot and milling machine*

parsing has been completed. The message (M-3) then pursues this course, getting parsed and processed by the appropriate specialist. Finally, the specialist sends back its new state information (M-4) to the **Cell Manager**.

As a result, new rules are fired, and new command messages (M-5) are generated for the next step in the cell's operation. These messages are also directed through the system dispatcher, and finally result in action being taken by a machine specialist (either **Robot-1** or **Milling-1**). The machine specialist builds a command suitable for an external device and sends it out (M-6). Finally, the protocol handler encodes it for transmission and sends it over the network for execution. In this example, the message (M-1) could have come from a robot that is providing information about the stock part to a milling machine. The intervening messages are internal status messages, intermediate and final translations.

$Lang

$Lang	Parser	Evaluator	Workspace
Robot	VAL	Interpret	Robot-1
Vision	RAIL	Interpret	Vision-1
Milling	NC	Interpret	Milling-1

Figure 4.15 *Some of the information in the system dispatch workspace*

This array of activity utilizes three principal ideas to maximize the flexibility of the resulting control system.

- *Events.* CML is event driven. Commands and messages can be sent at any time, without fear of them being lost or mishandled. They are processed in a priority-ordered first-in–first-out scheme.
- *Agenda system.* CML maintains a list of its activities and can make decisions about when and where they should be processed. For example, in critical periods it is important not to start a job that requires too much effort: during a manufacturing run, it would be wise not to start preventive maintenance.
- *Rules of interaction.* The cell manager is primarily concerned with the interaction of the machines. When a message is received from a device specialist, the state of the specialist is updated and the rules are scanned.

This software architecture has proven to be robust and is flexible enough to handle all sorts of situations. The next section concentrates on how a system of this sort can be built with a minimum amount of effort in CML.

6   TEACHING COOPERATIVE ACTIONS

Teaching sequences of fixed actions to robots has been standard fare since the robot revolution of the late 1970s. However, this approach of *teaching by doing* has never been successfully extended to more complex conditional sequences (i.e. programs) that are necessary to represent complete robot programs and multiple machine interactions. The solution to this problem is apparent within the context of a CML environment.

Robots cannot move inside a furnace without opening the door. Furnace doors must be closed as robots depart, or hydraulic hoses will melt. To avert potential catastrophes such as these, the machines must be carefully interlocked. This interlocking is accomplished in CML by using rules that limit the conditions under which actions can occur.

There are several approaches to the generation of new rules. The first approach is the obvious one: think of and write all of the necessary conditions for each action. While this approach is tedious, and often quite difficult, it does have two advantages.

(1) The rules can be concisely constructed to be logically minimal.
(2) The rules are not temporally ordered, and fire whenever an action is possible.

Unfortunately, manufacturing cells are often assembled by non-programmers who are not accustomed to building logic-based assertions. To avoid catastrophic errors, CML provides a second, automatic approach to the development of these rules.

This second approach allows the rules to be taught by using simple graphics and pointing (e.g. mouse technology). Each machine is represented by an icon which can be referred to by a pointing action. When there is a point in execution where no rules are found to execute, then a person is given the chance to add a new rule for the situation. The rule is constructed from the current system's state (Figure 4.11) and can be paraphrased as:

if *the-cell-is-in-this-state* **then** *proceed-with-indicated-action*

and which, when written out as a pattern-directed function call (Figure 4.16), shows that the rule is just a simple transformation on the states table. This rule is then automatically added to the rule set. This rule set is matched to the **Cell-States** table, and if no rules are fired then the otherwise clause is executed (**Build-New-Rule** in Figure 4.17). However, since a rule was just added to match explicitly the current state configuration, it will fire. This function results in the machine advancing to its next state and then the rules are checked again. This time if a rule is found, then the cell continues execution, and if not the operator points, a new rule is built, and execution continues.

If each state of the cell was equally likely, then this approach would not be very effective, because the number of rules would explode exponentially. Fortunately, this simple algorithm generates the key rules after only a few cycles. In our experience, it took about four cycles in a cell making one part style with nine machines and one bottleneck (i.e. the swaging process). If the range of parts is increased, then the timing variations of each manufacturing step start to increase. This increased variability causes new manufacturing situations that are not supported by rules; however, the teaching process can continue smoothly.

Rule

Rule	Eval	Func	Arg1	Val1	Arg2	Val2	Arg3	Val3
R1	eval	act	Robot	=Load-Furnace-R	Furnace	=Open-Door-C	Tool	=Machining-W

Figure 4.16 *New rule constructed from current cell state*

Eval-Loop

Eval-Loop	Command	Arg1	Arg2	Arg3
e1	while	~ExitLoop		
e2	expand	Cell-States2	Rules2	Build-New-Rule
e3	end			

Figure 4.17   *Top-level evaluation loop*

Since it is our goal to use this approach in small machine cells as well as in large flexible systems, it is useful to reduce automatically and sometimes weaken the generated rules. The process of rule reduction turns out to be simple, because it usually involves striking conditions from rules when the states in the machine's process plan have been covered. For example, suppose that a robot can be in two possible states {*is-working,is-waiting*}, and suppose that there are the following two rules:

> **if** *the-robot is-working* **then** *start-gage*

and

> **if** *the robot is-waiting* **then** *start-gage*

then these two rules can be reduced to one rule without conditions. Michalski (1980) reviews this inference rule and four others that effectively reduce the rule set, and Hayes-Roth and McDermott (1978) suggests a similar approach of rule abstraction. This example presents the weakest form of induction and amounts to nothing more than logical equivalence. However, there are other less stringent ways of reducing rule sets. For example, meta-rules can be added to the system that determine when rule conditions can be stricken, thus weakening the rules and broadening the conditions under which they apply. We have experimented with a number of these meta-rules. The simplest one removes conditions and reduces rules when a majority of a state domain has been covered. For example, if we extend the state domain for the robot to three states {*is-waiting,is-working,is-idle*} then this meta-rule allows the same inference we have already made from the given rule set, thus assuming that it is safe to gauge when the robot is idle. Unfortunately, this rule can be quite risky, because the inferred rules cannot be considered safe for critical machine interactions.

There are safer heuristics for inductive inference. In physical systems, such as in manufacturing, there are physical spheres of influence. In other words, machines that are in reach of one another must be collectively programmed more conservatively than machines that are geographically far apart. This forms a rather natural two-tier rule system, where the first layer of rules considers the geographical reachability between machines, and only then does the second layer of rules consider the relative machine behaviours.

Teaching these reachability rules can also be accomplished within a graphics-oriented system. A simple approach is to draw concentric circles around a machine until the scope of each machine's influence has been properly circumscribed. This data also provides information about other aspects of the system; for example, it can be deduced that machines with a broad reach are probably being used for material transport.

We are continuing research in this area, extending these ideas to larger flexible manufacturing and assembly systems. In order to avoid the exponential rule explosion, these larger systems demand heuristics for selecting and combining rules. Fortunately, many of the required heuristics are not needed for machine interactions, but are needed for planning factory schedules.

7   INSTALLATION OF AI IN THE FACTORY

It is difficult to install any advanced technology in the factory today unless there is at least one trained specialist. These 'advanced' technologies pose problems just because they are advanced and rarely, if ever, can be purchased as turnkey systems. This has led us to introduce to the factory AI tools that are designed to increase the productivity of system builders. The long-term goal is to make it possible for non-programmers to manage large software projects.

Figure 4.18 illustrates an array of these tools that have made it possible for non-programmers to develop the control system for a complex manufacturing cell. The 'y-axis' of Figure 4.18 describes how the CML tools are used. The automatic tools will generate and understand programs that fit with a defined family. On the other end of the y-axis, there are tools that aid the engineer in the remaining programming tasks. First, some of the programs can be 'taught' instead of programmed; and second, a special tool has been built that is itself an expert in CML and which gives advice on how to program.

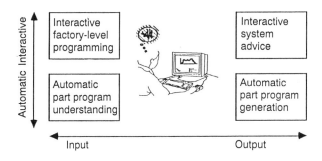

Figure 4.18   *AI tools for increasing software productivity*

The 'CML expert' takes a CML program as input and then writes a letter to the programmer. This letter recommends program optimizations, comments on program style, gives warnings about obsolete functions and gives advice about program structure. This kind of tool is very important in the manufacturing environment, since it cannot be assumed that the users are proficient in programming. Figure 4.19 gives an example of a letter written by the CML expert.

This letter is made from a series of preprogrammed paragraphs. These paragraphs are invoked using a rule-based system in CML and are accumulated in an outline. Once the first selection of paragraphs is made, the outline is scanned and modified so that redundant paragraphs are removed and appropriate concluding paragraphs added. Hopefully, the final letter can then be used by the programmer as a guide for improving the program. This simple approach to letter generation is being extended to independent sentence and subsentence generation, so that the criticisms of a CML program can be uniquely tailored to pinpoint its most basic problems.

8   SUMMARY

The cell management language provides computational tools that make it easy to: build a system of interpreters, program automatically solutions to manufacturing problems, and cope with communication problems that are prevalent in manufacturing systems.

We have set the goal of integrating new factory systems without using any traditional programming (i.e. typewritten logic). This remains an ideal, but we have successfully realized a significant reduction in the programming time for these systems. The programming that remains will be further reduced as we accumulate a large library of device interpreters. Much of the traditional programming is alleviated by a convenient teaching environment, written in CML, and the ability to reprogram automatically new solutions for controlling irregular situations.

9   REMAINING WORK

Once machines are properly connected and are cooperating in manufacturing activities, the weak link in manufacturing becomes the individual machine. We observed this time and again in the Westinghouse manufacturing cell, where an error would bring a single machine to a halt, which in turn would effectively bring the entire system to a halt. The problem was that the individual machines were not smart enough to detect and describe an error to a centralized CML decision maker, which makes it impossible for the system as a whole to make an intelligent plan for recovery.

Jun 21, 1989

Dear David:

I have spent some time analyzing the functions you have used in CML,
and have come to a number of conclusions. Nothing I say here should
be taken as absolute, since the analysis of your program has only been
performed at a surface level.

A huge portion of your program is devoted to input and output
operations. Perhaps you should try to centralize some of these
activities in one module.

You have used quite a few different control features, which indicates to
me that you have spent quite a bit of time refining the program.

You have used a number of database oriented operations, which tells me
that you have appreciated at least this aspect of CML.

I have noticed that you tend to use quite a few begin-condition commands.
It is usually quite easy to bring these together into a set of rules that
will show the logical structure of a program at a glance. The other
advantages of using rules is that they tend to run faster than their
"beginc" counterparts, it is often easier to change the program to
manage more situations, and to include parameters which results in a smaller
program.

Input statements are often convenient to use if you are writing a
"test" program, but they should not be part of any production software.
The problem with using the "input" statement is that it does not verify
that the input-string is an appropriate response. Try to use the "connect"
statement and parsing instead.

I noticed that you are using a "print" command. These are often
convenient in test programs, just as a quick way to write out some sort
of message. However, you should seriously consider switching these to
generate commands, since the output of a "generate" command can be
easily redirected to external processes and files.

Thank you for trying the CML expert system, I hope my comments about
your program have proven to be helpful.

          Sincerely,

          The CML Wizard

Figure 4.19   *Letter automatically written by CML expert*

On one occasion in the Westinghouse cell, each preform that was being swaged was becoming slightly longer than the previous one. Eventually, it became obvious to outside observers that one of the next parts would not have adequate clearance for the robot's unloading operation. Neither the robot, nor the open-die forge had a sensor to detect such a problem. Furthermore, while we could push emergency stop buttons on the forge and robot, that would cause new problems: the robot would drop the 1500 °C, 140 pound part after the hydraulic power had been turned off. In this case, the human operator had to grab the part with a large pair of tongs before it fell to the ground. After the situation was stabilized, all of the machines had to be carefully reset and synchronized with the CML hub so that it knew what actions had gone on behind its back. Minimally, this synchronization involved the machines sending their new state information to the CML hub after any human intervention.

As can be imagined from this example, one small error can easily start a chain reaction of errors. In retrospect, the first error is rarely serious in and of itself, but unrecognized it can quickly lead to bigger and much more serious problems.

9.1   Challenge for the 1990s

Automating normal operations, even in complex tasks, is slowly becoming routine. However, when errors strike, technology fails. The research of the 1990s must focus on how to recognize unforeseen errors and then manage them to minimize their impact.

10   ACKNOWLEDGEMENTS

This project has only been possible because of generous grants and support from the Westinghouse Electric Corporation. At the Westinghouse Turbine Components Plant, I would especially like to thank Jerry Colyer, the project manager, for his long-term efforts and interest in CML.

REFERENCES

Bourne D. A. CML: A meta-interpreter and manufacturing. *AI Mag.* 7(4), 86–96.
Bourne D. A. and Fox, M. S. (1984). Autonomous manufacturing: automating the job shop. *Computer*, **59** (243), 76–88.
Fussell P., Wright P. K. and Bourne D. A. (1984). A design of a controller as a component of a robotic manufacturing system. *J. Manuf. Syst.* **3** (1), 1–11.
Hayes-Roth F. and McDermott J. (1978). An interference matching technique for inducing abstractions. *Commun. ACM*, **21** (5), 401–10.
Michalski R. S. (1980). Pattern recognition as rule-guided inductive inference. *IEEE Pattern Anal. Mach. Intell.*, PAMI-4, 349–61.

Simpson J. A., Hocken R. K. and Albus J. S. (1984). The Automated Manufacturing Research Facility of the National Bureau of Standards. *J. Manuf. Syst.*, **1** (1), 17–32.

Veilleux R. F. and Petro L. W. (eds) (1988). *Tool and Manufacturing Handbook*. Dearborn, MI: Society of Manufacturing Engineers.

Wallstein R. S. (ed.) (1985). *CML Reference Guide*. Robotics Institute, Carnegie Mellon University (available from the author on request).

Wright P. K. and Bourne D. A. (1988). *Manufacturing Intelligence*. Reading, MA: Addison-Wesley.

Wright P. K. *et al.* (1982). A flexible manufacturing cell for swaging. *Mech. Eng.* **104** (10), 76–83.

5   A manufacturing controller implementation environment and its application at Rolls-Royce Sunderland

David Burnage and Terry Jones

1   INTRODUCTION

This chapter presents an advanced commercial product, CIMPICS (Computer-integrated Manufacturing using Pictures), the object of which is to allow the rapid construction of cell and area controller environments for a range of manufacturing applications. CIMPICS does not attempt to provide a single turnkey solution to all integration problems; it provides an environment to allow efficient customization of generic tools to particular problems. The product itself is based around both a suite of software tools that have been specifically configured to allow control activities to be easily specified and implemented and a range of utilities that ease the process of integrating and interfacing a wide variety of different manufacturing devices from different vendors.

The CIMPICS product represents the results of at least 75 man years of software effort, developed with the partial support of the UK Department on Trade and Industry. The original decision was made to develop the product after several years of experience of developing cell control solutions for customers. This taught Reflex that the key to progress in this area was the development of a set of software tools that allowed the company and its customers to integrate and control devices from the customer companies' preferred hardware vendors and existing manufacturing plant, this all being carried out within the real manufacturing constraints associated with non-green-field installations.

The development within Reflex built upon a philosophy of adopting well-understood techniques such as relational databases and, wherever possible, emerging standards including Unix and C, MAP, and sophisticated graphical tools and user interfaces such as X Windows (Scheifler and Gettys, 1987) and Grafcet (IEC 848, 1988).

Key objectives of the development were that the final system should be as computer-vendor independent as possible, enable migration to more powerful computers as necessary and provide users of the system with tools to reconfigure, amend and extend the system as their needs changed.

The software system allows the user to integrate plant elements that may at present be functioning as 'islands of automation' with the minimum amount of support from computing professionals. The integration approach implicit in CIMPICS was initially intended for bottom-up integration of systems but the increasing recognition of the need to control and schedule semi-automated systems is leading to equally successful top-down integration projects.

This chapter begins by introducing the product and continues by describing the graphical tools within the product that are used to generate control solutions. The chapter closes by indicating the form of cell and area controllers that are conventionally built within CIMPICS.

2   CIMPICS

The system constructed is a modular system, running under UNIX, that can be configured to suit any level of factory automation. Based on a single common database, it allows a user to start with a single manufacturing cell and expand the application to include areas and ultimately the factory. CIMPICS is intended to integrate with industry-standard process control computer systems.

Central to CIMPICS is its graphics-based interface. Reflex have created a graphics interface particularly targeted at CIM – the CIM workstation. Using standard windows, icons, menus and pointers, the CIMPICS man–machine interface provides convenient, easily understood access to the system. This is the main interface into the system for configuring cell controllers, planning the manufacture of parts through them, and monitoring and debugging cell operations.

In contrast to producing a control system that requires extensive customization for each manufacturing site, controllers are constructed from a range of building block system tools within CIMPICS. Using the graphics interface, the system implementation is performed by selecting the required features, links and procedures from standard menus. This also allows simple subsequent system modification by the user.

In order to integrate devices into cells it is necessary to use the real-time database (RTDB) of CIMPICS. All cell controller user data is located in the RTDB, a simple-to-use shared memory facility. Any task within a cell controller is able to access the RTDB. For critical operations data access is possible in under 300 microseconds, and a number of locking options are available to enable the application software to lock individual items or whole sections of the database. The RTDB also allows for partitioning of the database, and provides debug tools and extensive facilities for error recovery.

As it is necessary for CIMPICS to integrate a variety of automated processes which utilize a wide range of different operating systems, interfaces and protocol, Reflex has adopted the Manufacturing Automation

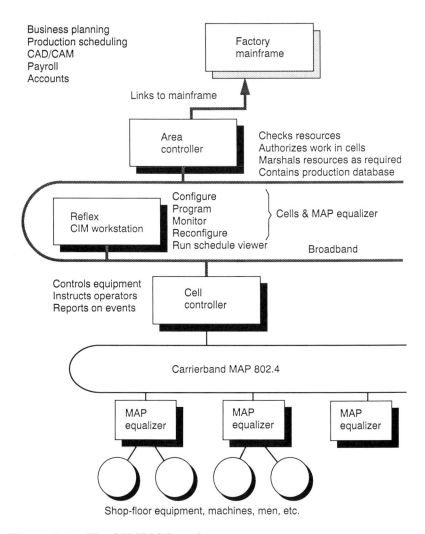

Business planning
Production scheduling
CAD/CAM
Payroll
Accounts

Factory
mainframe

Links to mainframe

Area
controller

Checks resources
Authorizes work in cells
Marshals resources as required
Contains production database

Reflex
CIM workstation

Configure
Program
Monitor
Reconfigure
Run schedule viewer

⎫
⎬ Cells & MAP equalizer
⎭

Broadband

Controls equipment
Instructs operators
Reports on events

Cell
controller

Carrierband MAP 802.4

MAP
equalizer

MAP
equalizer

MAP
equalizer

Shop-floor equipment, machines, men, etc.

Figure 5.1   *The CIMPICS products*

Protocol (MAP) to allow interfacing and communication between processes. This includes the Manufacturing Message Service (MMS) and File
Transfer, Access and Management (FTAM) protocols, and is supported by
the Inter-Process Communications (IPC) mechanism within CIMPICS for
direct process-to-process communications.

CIMPICS has been designed to accommodate the complete hierarchy, as
is shown in Figure 5.1, that can exist within manufacturing industry. The
highest level may represent an entire organization including a number of
separate plants. Within any plant, the area controllers and cell controllers
will be of many different sizes to handle the number and complexity of the
different applications. A cell may contain a single machine tool, or a group of

several machines and other supporting equipment. Areas may be large or small, and functions such as tool management, quality management and capacity management will vary enormously in complexity between different manufacturers. To cater for this variety of requirements, the CIMPICS modules and the system links and processes are defined by function, not by plant size or number of machines.

CIMPICS has specifically been configured to allow the construction of strictly hierarchical control systems. In CIMPICS there is no direct peer-to-peer communication between devices on the same level in the manufacturing hierarchy. This is in direct contrast to other research-based systems discussed in this book. With the present generation of manufacturing control processors a hierarchical approach does, however, give a good price performance for supervisory control without an undue communications overhead.

We now begin our detailed discussion of the product by examining the CIM workstation.

3　THE CIM WORKSTATION

This has been designed to be the main interface between the user and the CIMPICS system and is an advanced 'toolbox', containing pictorial system design and programming tools, debug tools, and graphical cell monitoring and fault-tracing tools. The CIM workstation presently runs on a SUN workstation, but is easily ported to other manufacturers' platforms which are suited to the graphical display of information.

The CIM workstation makes use of windows, icons, mouse and pointers and the X-Windows device-independent display system. The interactive graphics development tools are based on the standard Grafcet control programming system, which is used to specify the relationships between processes. Within each process the CIM workstation uses a powerful graphical programming language, Reflex Graflex, to program the individual actions to be carried out in a single process. Graflex uses simple user-defined symbols to represent the components described in the process which is to be controlled. The combination of Grafcet and Graflex allows engineers – who are not necessarily experienced computer programmers – to write applications programs on the CIM workstation for execution in the cell controller. Figure 5.2 shows the suite of tools available within the CIM workstation.

The main functions of the CIM workstation are configuration, programming and management of the cell controllers, as well as cell monitoring and fault tracing. The schedule viewer software also resides in the CIM workstation and uses the output of the area controller – the 'work-to-list' – as its input. The schedule viewer presents the area schedule in a Gantt chart style, on a 'per machine' basis, against a timescale measured in hours, shifts or days, as selected. As work progresses, some area rescheduling may be

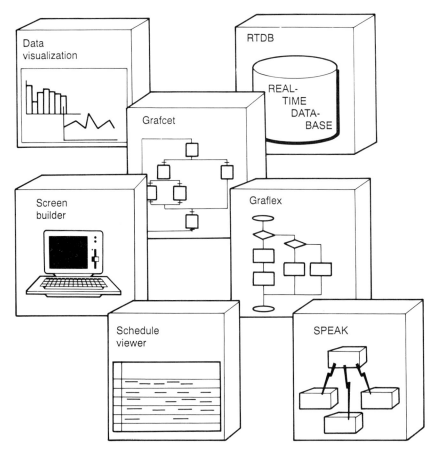

Figure 5.2 *The CIMPICS software tools*

required to take account of machine breakdowns, staff absence or changes in resource availability. The updated area schedule is relayed automatically to the CIM workstation to keep the schedule viewer completely up to date. These features will be described subsequently.

Developments of the CIMPICS system are increasingly expanding the role of the CIM workstation to include configuration, programming and management of higher-level area controllers.

The controllers constructed using CIMPICS are intended to run on a variety of UNIX-based workstations. Cell controllers that require real-time control decisions require machines with operating systems that respond in real time and that may allow multi-tasking. At present these controllers are constructed using VME-based solutions supporting the proprietory real-time operating systems MTOS ™ and OS9.

We now turn to examine the key features of the CIM workstation, beginning with the graphical programming languages.

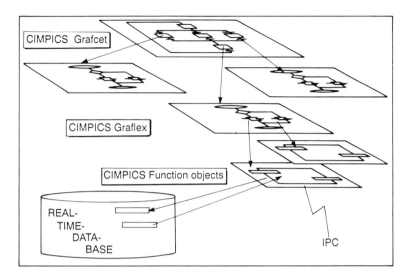

CIMPICS Grafcet

CIMPICS Graflex

CIMPICS Function objects

REAL-
TIME-
DATA-
BASE

IPC

Figure 5.3 *Graphical programming environment*

4   GRAFCET/GRAFLEX

Grafcet/Graflex is a graphical programming environment for the preparation of application programs on the CIM workstation for execution in the cell controller. An overview of the elements of the programming environment is shown in Figure 5.3. The symbols within the language allow the design of complex CIM systems, the programming of concurrent control systems, and the control of synchronous and asynchronous processes. Graflex is a Reflex product that extends the functionality of the control language Grafcet.

Graflex is a graphical language that allows function objects to be logically linked with flowchart-like sequential control structures. A function object, the base-level utility of the language, is a precompiled C routine that can, for example, access the real-time database to extract and amend data or, for instance, send a message to another function object.

Graflex behaves as the programming level between Grafcet and the function objects. The purpose of Grafcet is to specify graphically the relationship between processes programmed in Graflex, as is indicated at the top of Figure 5.3. Grafcet is cyclic in nature and program steps are repeatedly executed until their associated transition becomes true. More than one step can be active at any one time allowing concurrency.

Graflex is used to program the individual actions to be carried out in a single process, as is indicated in the centre of Figure 5.3. Graflex executes from the top down, as a single thread. It has no wait states, but can test whether an event has occurred and return TRUE or FALSE to the Grafcet transition. For this reason Graflex has no do–while or do–until loops.

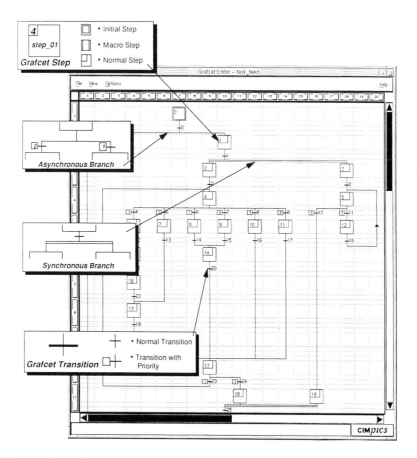

Figure 5.4 *Grafcet programming*

Grafcet programs can communicate via messages passed by function objects. A typical message exchange between Grafcet programs may take the following form: a message sent from one program could cause the receiving program, perhaps waiting at a transition for the message, to continue to the next step; the message content can be interrogated by the receiving function object and manipulated by Graflex programs.

Each Grafcet step segment is composed of up to three Graflex segments, one each for activation, continuation and deactivation. Each non-trivial transition segment is composed of a single Graflex segment that returns a value TRUE or FALSE to the Grafcet system.

The Grafcet/Graflex system is split between an editor on the CIM workstation and the run-time kernel running in the cell controller. A duplicate non-real-time kernel system is available in the workstation for debugging purposes.

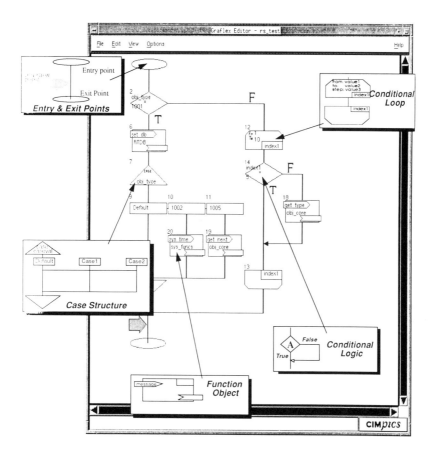

Figure 5.5 *Graflex programming*

Each Grafcet step represents a single process, and each process can be as simple or as complex as required. Logical progress from step to step is controlled by Grafcet transitions (each Grafcet step must, therefore, be followed by a Grafcet transition). Examples of Graflex and Grafcet programming instructions and structures are shown in Figures 5.4 and 5.5.

This structure therefore provides an object-like (Cox 86) environment for simplified program development and software reusability. Object-oriented facilities at present include encapsulation and data hiding and specialization, but there is no inheritance. Key software modules are supplied prepackaged, developed, integrated and tested. The use of software engineering techniques is encouraged by the automated application of structured software design techniques and standards, and the use of a universal flowcharting system and standard flowcharting symbols for the software design.

5 INTEGRATION UTILITIES

In an integration and control application, a number of utilities are required to allow the execution of the control logic that has been prepared using the graphical programming tools. These include a real-time database, a mechanism for allowing interprocess communication between control processes, and interfacing utilities.

5.1 Real-time Database (RTDB)

In an integration and control application which coordinates the activities of different devices it is necessary to use an RTDB.

In the Reflex database, all cell controller user data is located in the RTDB. Any task within a cell controller is able to access the RTDB, thus providing a simple-to-use shared memory facility. The user defines a data type only once within the RTDB system and is then able to 'export' that definition to the language system being used for coding. The type, offset and size of data items is available at run time. The offset only has meaning when the item is contained in a record. Related data may be grouped so that the same name can be used for data items which exist as separate entities in different contexts. These appear as a separate database, for example the same data names/items can occur for many different programmable logic controllers (PLCs) in the system but can be individually identified. The data item names and data are not compiled in the conventional sense or as a relational database requires. The data items may be created, inserted or deleted during run time. An application program needs to know only a database item name in order to access it. A method of mapping names to keys at run time for faster access is also provided.

The RTDB is structured as a schema with a hierarchical set of databases. It is accessed by a number of methods via a key (a safe access with a known pathname) or a 'preget' key (a particular named key that allows a faster method of accessing particular identified data).

For database error recovery, if the RTDB detects an internal error it will return an error code to the caller. Also, if an application needs to recover from an error it can use the initialize reset facility to tell the RTDB to reset a database section. This causes all items which have a reset value to be reinitialized to that value.

5.2 Interprocess Communication (IPC)

The Reflex IPC provides two services to the user: First, it assists the user process to obtain a connection to a service; and second, communication is provided between those processes once a connection is established.

Whilst data transport may be provided using either an ISO standard network or a proprietary network, or indeed by the underlying operating system, the objective of the IPC is to provide a uniform interface for software applications across all machines, operating systems and networks throughout the CIM environment. This is achieved by providing a standard functional interface on each machine and a library for use with each supported programming language, such as C and Grafcet/Graflex. The benefit of this facility is that software can be rapidly reconfigured or migrated to other areas of the system. For example, software written initially for the area controller may be easily migrated to the cell controller or vice versa.

IPC supports a server–client model. A process offering a service will open one or more channels as a server. This action notifies a directory service that the server is active. An open request from a client process scans the directory for a service of the requested type and establishes a channel. Services can be unidirectional or bidirectional and the open requests from each party must match in this respect for the requests to be successful. Servers may offer more than one service or the same service more than once. In the event that more than one service is available to a requesting process, the first available service is connected.

Communications are message oriented. A message is accepted as one unit at the transmitting process and delivered as one unit at the receiving process. Message transfer is fully buffered so that arriving messages are held until the destination process is ready to receive. The transmitting process is informed of the successful transmission. Status information is provided so that a process can check that the communications connection is functioning correctly.

5.3   MAP Equalizer

The integration of shop-floor devices requires the provision of utilities that allow messages to be passed both from the controller to the devices and from the device to the controller to indicate the status of the lower-level device. The MAP equalizer has been developed to satisfy the requirements for such an industrialized shop-floor data collection unit, for use below a cell controller. This hardware device will support both serial and digital I/O with a variety of interfaces for each, for example MAP 3.0 Carrier Band (802.4), Serial I/O (RS232, RS422, Current loop) and digital I/O. The interface between the MAP equalizer and its host machine conforms to the MMS standard using the MMS variable-access services. The device allows extension of its capabilities by downloading additional programs, using MMS domain management services. As the MAP equalizer has a powerful CPU and a large amount of free memory, it can be further enhanced using additional software tools.

5.4   SPEAK (Serial Protocol Editor and Kernel)

One of these tools is SPEAK, the Serial Protocol Editor and Kernel. In general the Reflex CIMPICS project has adopted the MAP and TOP specifications to provide the underlying communications framework. In particular, the use of the MMS specification in MAP provides a standard set of services to enable communication with shop-floor devices. Unfortunately, most of the devices currently installed in manufacturing sites neither have a MAP or a TOP interface nor could be upgraded to include one.

Many devices, however, do support simple serial communications combined with a link-level-type protocol. The functionality provided by the proprietary protocol is usually sufficient to support simple communications tasks such as upload/download, status reporting and error reporting. There are no generally accepted standards for the link-level protocols used and so it is costly for the end user to support the diversity of proprietary protocols found in many current manufacturing sites.

The objective of SPEAK is to enable link-level handlers to be developed rapidly and efficiently to support serial communications. The SPEAK toolkit provides the user with an integrated development environment for serial protocol implementations.

This window-based toolkit provides a flexible set of modules to allow the user to specify, develop, implement and test the serial protocol for communications to a device using the serial protocol. The user specifies the serial protocol using a graphical editor, which produces a protocol-independent specification format. This specification is then used to generate the implementation. The protocol is modelled as a finite-state machine (FSM) which is represented by a state transition table. For each state there is a row, and a column for each event. Each cell (intersection of a row and column) of the table is therefore a transition state and triggers an action routine. The state transition table must be accompanied by a specification of the event conditions, action routines and the FSM variables: the event condition is an algorithm for determining if an event has occurred, the action routine is performed when an event occurs, and the FSM variables are variables used within the protocol. The graphical editor gives the user a simple-to-use interface for specifying the state table of the FSM which represents the protocol. The toolkit automatically invokes the editor, for the user to write the actions and event condition routines.

The toolkit includes a library of frequently used event and action routines, which further simplify the task of building the implementation. The specification generated using the editor can then be checked for consistency. The consistency checker will generate a report giving warnings where the state table specification may cause problems for a protocol implementation. For example, a state which has no corresponding transition to another state is valid within a general FSM, but within a protocol this is not sensible as it would cause a deadlock in communications. The user can then update the specification using the graphical editor.

Once the protocol has been fully specified and all the action and condition routines written, the user can generate the implementation. The generator will combine library routines, the state table information and the user's code to produce an implementation. The implementation can optionally generate a debug 'trail' for later analysis. An implementation with debug code will produce a log file while running, which can then be used with the debug display unit to animate the operation of the protocol graphically. The information from the debug log may highlight deficiencies at the specification phase, indicating the user must continue the development cycle.

6   CIMPICS APPLICATIONS

The purpose of the tools described above is to develop controllers, and this section of the chapter describes two of the controller environments generated by Reflex using the toolkit.

6.1   The Area Controller

The number of area controllers in a system will depend on the size and complexity of the plant. The main function of an area controller is to receive the works orders from the plant control system and to schedule work-to-lists for each manufacturing cell. The area controllers operate on a shorter timescale than the plant controller. Typically, this range covers a time horizon of the weekly production schedule, and a coordination timescale that corresponds to a production shift.

The principal functions performed by the area controller are as follows.

6.1.1   Work scheduling. On receipt of the works order, the work-to-list, the area controller reconciles the new work-to-list with the current one and checks the resources required and those available. It then reports that the requirements either can or cannot be satisfied. If the requirements can be satisfied it then distributes the work between the cells and coordinates the resource allocation scheduling and rescheduling as necessary.

6.1.2   Marshalling. The requisition of materials and marshalling of stores, tools and manufacturing data is all handled by the area controller. Requisitions from stores are linked directly with stock control, statistical and quality control subsystems. This system also marshals manufacturing data including CNC part programs and performs part tracking.

6.1.3   Transport and storage. The area controller has facilities for managing both automatic and manual transport systems.

6.1.4   Personnel control. By communicating with plant-level employee availability control, each area controller manages the available personnel resources to achieve the required schedule. This activity includes recording personnel time and attendance.

6.1.5   Maintenance management. Both planned and unplanned maintenance are managed by the area controller for the equipment within the area, including breakdown information and fault logs to give cell uptime.

6.1.6   Cell management. The central purpose of all the scheduling, marshalling and management functions is the process of generating work-to-lists for each manufacturing cell and the transmission of detailed data to each work cell. This data includes cell programs, CNC data, robot programs and manual instructions. The area controller also maintains individual cell records, including cell performance, quality assurance and part tracking.

After receiving system feedback information from each of the cells which it controls, the area controller updates its database and reports to management the current situation, future trends, and a comparison with historical events.

6.2   The Cell Controller

The size and complexity of the manufacturing process will determine the number of machines in a manufacturing cell and the number of cells within an area controller. Each cell controller receives from its area controller the work-to-lists for the machines in the cell. The configuration of the cell is set up and maintained using the Reflex CIM workstation. Reconfiguring the cell to accommodate changed or new machinery can be handled by local production engineers, without requiring outside assistance.

The timescale for cell controller operations has a time horizon of a production shift, within which it operates under real-time control. The main functions of the cell controller are as follows.

6.2.1   Scheduling cell operations. The cell controller receives the work-to-list and the manufacturing data from the area controller, schedules the individual cell operations for each machine, and reports 'cannot do' situations.

6.2.2   Generating timescales for each machine. The cell controller collects machine, equipment and operator status, and sequences components through machines. By linking operations on components, the cell controller can reduce waiting times and batch queues to make savings on both work-in-progress and overall lead times.

6.2.3   Marshalling. At a detailed level, the cell controller marshals both local and external instructions, materials and tools to satisfy the requirements of each machine. As it has constant online access to resource availability data on the area controller, the cell controller can reduce downtime due to unavailable resources.

6.2.4   Initiation and control of manufacturing. The cell controller supervises the control and coordination of machines, operators, materials, tools and information by sending commands to the manufacturing resources. It provides an unmanned operation capability, controls automatic interlocks and safety guards, includes a direct numerical control facility (DNC) and coordinates manual activities. The cell controller also provides for external coordination of transport devices, both automatic and manual, and dispatches. Dispatched items include finished and rejected components, used tools and their life status, and unused material and waste. It can also coordinate quality assurance by collecting inspection information and statistical data, and it can interface to coordinate measuring machines.

6.2.5   System feedback. By receiving information from each machine controller via digital I/O or manual operator, the cell controller constantly updates the system database by reporting to the area controller complete details of progress, problems, part tracking, and operator time and attendance.

6.3   The Schedule Viewer

One of the primary functions of the controllers is the generation of schedules. A key feature of the CIMPICS toolkit is that it allows the schedule generated to be reviewed. MRP production schedules have historically been produced by computers overnight, in a batch run mode, and printed on to large volumes of computer paper. Foremen, supervisors, works managers and progress chasers subsequently pore over these documents for the following shift/week/month in an effort to manufacture that which the computer orders. Inevitably, as time and work progress, the planned schedule diverges from reality as 'red-hot' jobs intrude and batches are split for 'rush jobs'. The original printout becomes out of date or covered with notes and anecdotes. In either case, it is difficult to assimilate the very important information it contains. What is required is an easy-to-understand view of the production schedule which can be kept up to date as situations change. Columns and rows of works orders, batch numbers, part numbers and priority dates must be minimized and an alternative way of clearly depicting the schedule and the current situation devised.

The schedule viewer software resides in the CIM workstation and takes the output of the area controller scheduler – the work-to-list – as its input and generates such a view. The CIM workstation display is ideally suited to the graphical display of information. The schedule viewer presents the schedule in a Gantt chart style on a per cell or machine basis against a timescale measured in hours, shifts or days as selected, as shown in Figure 5.6

As the work progresses, the area controller may reschedule subsequent operations and batches to take account of machine breakdowns, staff absence or changes in resource availability. The new up-to-date schedule can then be displayed at the CIM workstation automatically. The schedule viewer display may be arranged in a number of ways to aid the understanding of the shop supervision. For instance, the display may show a group of

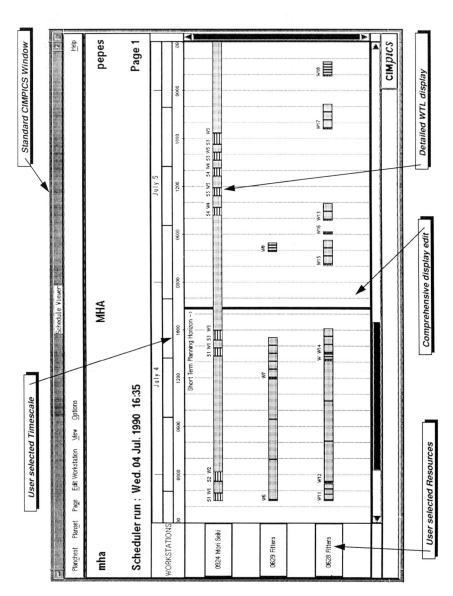

Figure 5.6 *Schedule viewer system*

machines and all the components in progress, or it may be arranged to show the path of a single component as it passes through a number of set-ups and operations on different machines. Display configurations are performed by the user using the standard graphical tools.

The CIMPICS area controller takes as its input the plant-generated production schedule and combines this with its detailed knowledge of the current manufacturing status. It then performs a much more detailed and refined scheduling process over a short period of time. The output of the area scheduler is a work-to-list which can be performed by the available machines and a list of components which cannot be manufactured for a variety of reasons – the 'can't-do-list'. The 'work-to-list' is passed to the cells over which the area controller has jurisdiction. The area scheduler can be rerun on a much more regular basis and stays up to date with progress on the shop floor by taking feedback information from operators, machines and supervision, and manages updates and changes to the original production schedule.

The present scheduling implementations take the order priority set by an MRP system, and run using data collected from the manufacturing environment (at, for example, 30 minute intervals) to allocate the highest priority works orders to machines that are free. The start times for these jobs and their necessary resources are then set by the area controllers at, typically, half-shift intervals. A separate procedure then organizes the transport tasks required to marshal the necessary manufacturing equipment, the data being passed to this procedure via the cell controller from the area controller.

We now turn to describe a particular installation of the CIMPICS product.

7   CASE STUDY

7.1   Rolls-Royce – The Company

Rolls-Royce is one of the United Kingdom's foremost engineering companies employing the most advanced technologies. It has a skilled workforce of 64 900 people and is one of the country's biggest exporters with 75 per cent of Group sales being for overseas markets.

Rolls-Royce designs, develops, manufactures and supports gas turbine engines and ancillary equipment for aircraft and for industrial and marine applications. Its aeroengines operate up to the highest thrusts for civil and military aircraft. Its power plants are used by over 300 airlines, 700 executive and corporate operators and over 100 armed services throughout the world. Nearly 200 industrial customers use Rolls-Royce gas turbines for power generation, gas and oil pumping and other industrial purposes. Naval vessels from 25 countries are powered by Rolls-Royce engines.

Through NEI, the Group is engaged in the design, manufacture, construction, commissioning and servicing of capital plant and equipment for a wide range of industries, with a particularly strong emphasis on power generation. NEI operates worldwide in both manufacturing and project management.

Rolls-Royce and NEI merged in the spring of 1989, which represented the first significant diversification for Rolls-Royce in that 35 per cent of its turnover in a full year is now expected to be in products other than aero gas turbines.

NEI is beginning to benefit from Rolls-Royce's unrivalled experience in high-technology manufacturing and extensive investment in research. Cost reductions are being achieved overall by combining operations, functions and offices together.

7.2   Rolls-Royce Sunderland

Rolls-Royce's site in Sunderland, the Pallion Site, overlooks the River Wear and employs 640 people.

Bristol Aeroplane Company opened the Site in 1952 initially making pistons, sleeve cranks, main drive shafts and gears. Within 2 years of that date, it was making major components for Proteus, Orion, Ramjet, Pegasus and Olympus engines.

During the course of the 1980s the DTI gave financial aid to Rolls-Royce to build its new main factory to house the disc manufacturing cell. There are now four integrated factories covering 5 acres and a £4.5 million investment has been made in CNC and DNC machine tools.

In the aerospace industry a most critical component in an engine is the rotating disc that holds the compressor, turbine and fan blades. Rolls-Royce in Sunderland currently produces 60 per cent of these critical high-value components. They are made from titanium-, steel- or nickel-based alloys and cost from £2000 to £25 000 each.

Over the last 2 years the Sunderland facility has consolidated its strategy on component manufacture and continued its investment in modern machine tools and systems. The year 1989 saw the opening of the new fan disc cell which will ultimately be capable of producing all the company's fan discs.

The site in Sunderland stands above the remnants of Wearside's shipyards. It is a poignant reminder of the changing face of industry. Shipbuilding was once a measure of a country's international prestige; that position is now arguably held by aerospace. Rolls-Royce's Sunderland plant is at the forefront of manufacturing technology which will keep it ahead of the competition.

Aerospace is a global market and the use of technology is reflected within the industry. It has been described as an industry of apparent contradiction, competing for international business with one hand and collaborating with the same companies with the other.

It is inevitable that all aerospace companies would look at open systems to improve communications – Rolls-Royce is no different, as can be seen. The aim of manufacturing is to become more efficient, not less. As well as needing open systems to communicate with collaborative partners, Rolls-Royce needs to communicate with a multitude of suppliers, many of whom are involved in shared risk and revenue agreements. Open systems in isolation is not a solution. It frees the company from restrictions in the choice of hardware, software, networking and communications, but the competitive edge for the 1990s and beyond must come from changes in manufacturing systems to support local and corporate needs.

In manufacturing terms this means cutting lead times, reducing inventory and generally rationalizing operations. The business at Sunderland might be aerospace, but priorities there are the same as any manufacturing company. Sunderland is seen as a 'standalone' unit with a massive turnover, employing many people. Rolls-Royce has recognized that autonomy can only be achieved if it moves away from centralized computer systems. In an environment where the goals are short leadtimes, increased flexibility is essential. To achieve the requirements necessary, production control has to be located where it can be reactive.

7.3   Supply Group

At the very heart of Rolls-Royce is the Supply Group. It has to produce everything that is needed to build an aeroengine from huge, high-value parts to the smallest nut and bolt – on time and at the right price. The Supply Group is the company's biggest employer. Sixteen thousand people work in its nine major facilities in Great Britain. These factories have a floor area of 6 million square feet and house more than 4000 machine tools. The annual spend of the Supply Group is well in excess of £1 billion.

It is responsible for supplying 70 000 different parts. Of these, 20 000 important components are made in its own factories and the other 50 000 are manufactured for the Group.

The reorganization of manufacturing facilities into smaller shop-floor cells has been a significant advance during the latter part of the 1980s and will continue into the 1990s with enthusiasm. These shop-floor cells are reducing the lead times for component manufacture and cutting the investment in work in progress, thus permitting a faster response to customer requirements and a reduction in manufacturing costs.

Among the wide-scale changes taking place in the Rolls-Royce Supply Group are those in the vital area of manufacturing control. As well as keeping the process of progressing materials through the factories operating on a day-to-day basis, several initiatives are in hand to support the changes going on throughout the Supply Group. All these changes are aimed at making it world-competitive in terms of quality, on-time delivery, low inventories, short response times and low unit costs.

The control of a vast amount of materials and stock handling is an enormous challenge for the Supply Group. In a typical week £5 million worth of raw materials are delivered to various Rolls-Royce factories. This is subsequently launched into the £250 million worth of work-in-progress going through the manufacturing facilities. Nearly £20 million worth of deliveries to customers in the military and civil engine business groups is made each week from Rolls-Royce factories and external suppliers – hundreds of thousands of components spread across some 30 000 individual part numbers.

The logistical problems are of course enormous. If any of the deliveries are late, or if any operations are performed in the wrong priority sequence, this can result in deliveries to the business groups being deficient, thereby preventing the company from satisfying its customers. Similarly, if any deliveries are made early, the company is forced to spend money sooner than it needs, tying up capital in stock when it could be utilized for new projects or capital equipment. Stock moves every week from raw materials through each operation for eventual delivery to the business groups and involves many suppliers – more than thirty Rolls-Royce shops at nine major sites and hundreds of thousands of components.

7.4   Manufacturing Control Strategy

The objective of getting the right part at the right price at the right time, demands that all of the logistics associated with these details need to be performed in the most professional way. Getting the logistics right is the prime purpose of Manufacturing Control.

A number of strategic initiatives are taking place within Rolls-Royce to improve Manufacturing Control. The MRP (Material Requirements Planning) systems installed during the 1960s and 1970s now do not meet the needs of Rolls-Royce owing to changes in the market place, company changes and also an improvement in MRP systems. The company re-organization in the early 1980s created a single Supply Group, which effectively meant that the old site-based systems could not give effective support. The MRP system was designed to suit the major sites, the minor sites only being considered as an afterthought at the time.

Pressure to reduce lead times, both to the customer and as a means of avoiding holding large stocks of materials and parts, necessitated action to review the situation with regard to using modern technology to control all the sites strategically. The strategies employed to achieve these objectives are outlined below and enhance the manufacturing effectiveness of Rolls-Royce in providing a competitive advantage. This is where the careful consideration of the application of technology is so important; such systems must be applied in the context of overall systems and just-in-time philosophies through the manufacturing systems engineering (MSE) approach, which will be explained more fully later.

7.4.1   Strategies. The strategies are as follows:

- Domestic facility rationalization (product families)
- Facility redesign using MSE (cell manufacturing)
- Purchase strategy to integrate Rolls-Royce suppliers
- Logistics strategy (Merlin/MRP II)
- Totally integrated manufacturing engineering (TIMES)
- Improved shop control (Reflex)
- Simultaneous engineering (design and manufacturing).

To achieve the objectives of the Supply Group, these strategies had to 'come together' in a coherent way to maximize the efficiency of the Supply Group.

Considerable effort continues to be devoted to promoting the total quality approach throughout manufacturing within Rolls-Royce, and the total quality seminars which were conducted throughout the Supply Group achieved a high degree of personal commitment from all personnel involved. This has produced not only tangible improvements in the quality of work and further reductions in non-conformance levels, but also increased job satisfaction for employees.

7.5   Merlin

The MRP (Manufacturing Resource Planning) systems of the 1960s and 1970s, namely MAGPIE (Machine Automatically Generating Production Inventory Evaluation) and IBIS (Inventory Based Instruction System), are no longer suitable. However, with the marketplace changing and Rolls-Royce and MRP's techniques improving over the last two decades, advances had to be made to see Rolls-Royce's future as innovative and secure. Therefore, Rolls-Royce is moving towards the introduction of a replacement system known as MERLIN (Mechanized Evaluations of Resources, Logistics and Inventory). This will be the largest computer system project that Rolls-Royce has ever undertaken and will allow parts for any engine project to be ordered from any factory within the Supply Group, the older systems being site based rather than company based. This embodies the JIT (Just-in-Time) approach cutting large stock holdings and giving better, highly efficient use of resources.

7.6   Manufacturing Systems Engineering (MSE)

The MSE approach concentrates on the redesign of Rolls-Royce's factories by multi-disciplined task forces into cells created and designed to reduce lead times, inventories and non-conformance, which will be of enormous benefit to the customer as well as improving the quality of the work for employees by giving them more job satisfaction and a greater involvement in the industry. The follow-on effect will develop a total quality approach and a move towards just-in-time delivery.

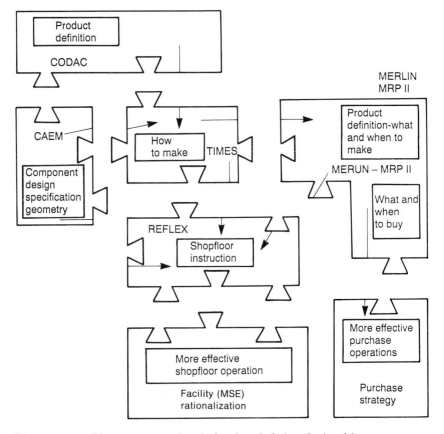

Figure 5.7   *The systems at Sunderland and their relationship*

7.6.1   MSE task force. A task force was set up and launched in 1989 after an extensive feasibility study was carried out to determine the benefits of MSE. The task force is led by specially trained multi-disciplinary leaders and also involves shop-floor employees. This team approach has also breached the barriers between design and production.

Experienced manufacturing engineers work closely with designers at the critical design stages, and as materials technology is such an important part of product design, and changes so rapidly, designers are now included in the materials development team. This ensures that design methods keep pace with materials development.

One of the greatest benefits of MSE is that existing plant and equipment is reorganized to make the best use of what is already there.

The task force works on a full-time basis and produce a plan to redesign the facility. A series of well-defined stages are gone through to reorganize work into cells, which results in an accountable structure with an emphasis on a total quality approach to manufacture.

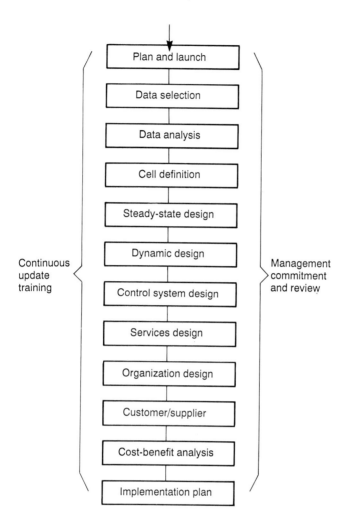

Figure 5.8 *The MSE design model*

MSE is a structured approach to the development of a more effective means of manufacture by

- improved use of existing resources
- the use of common sense and simple solutions to complex problems
- two-way communication with the whole workforce
- providing a mechanism for continuous improvement.

It also uses a 'bottom-up' approach in that it utilizes previously untapped resources by involving the whole workforce.

The MSE design model is shown in Figure 5.8.

7.7   The CIMPICS Collaboration

In December 1986, the Department of Trade and Industry (DTI) sponsored an exhibition to encourage British industry to look closely and carefully at computer-integrated manufacturing. Rolls-Royce participated in this exhibition, as did Reflex, the authors of the balance of this chapter. Subsequently, the DTI invited Reflex to form a consortium and encouraged it to take advantage of DTI funding under their 'Support for Innovation' scheme.

Rolls-Royce provided the user focus for this collaboration and also

- the user requirement specification
- project management and engineering at the Rolls-Royce Sunderland installation
- significant work on mainframe interfacing and data provision for CIMPICS, and
- the Sunderland Turbine Disc Cell Beta test site.

The software developed by Reflex as its part in the consortium now forms the basis for the CIMPICS product suite.

In the spring of 1988, Reflex became a wholly owned subsidiary of Rolls-Royce and, under the ownership and guidance of Rolls-Royce, it has continued to develop and invest in CIMPICS. A second version of the Sunderland system has been installed recently to replace the Collaboration Beta test version.

It is anticipated that CIMPICS will provide a significant contribution to the Rolls-Royce Manufacturing Control Architecture in the future. In addition, Reflex now provides the products in the open systems marketplace as an independent supplier.

7.8   Sunderland Requirements

The nature of the manufacturing within the Sunderland site, coupled with the findings of the MSE task force, made CIMPICS an ideal choice for complete CIM within the aerospace industry, although the application of CIMPICS is not restricted purely to the aerospace industry but can be utilized in many different industries.

The control of piece part manufacturing and the organization and methods to achieve efficient production have exercised the minds of production engineers for decades. The aim of the previous initiatives (i.e. MRP, FMS, MRP II, OPT and JIT) was to speed the flow of raw materials towards the creation of finished parts. In this situation, and especially in conventional machine shops with geographic/machine-type shop layout, the detailed scheduling workload is shouldered by supervisors, foremen, production controllers and chasers.

Outside pressures lead to real scheduling of only the highest-priority 'red-hot urgent' jobs. Other batches are fitted in as and when there is spare capacity, or when they have reached a degree of lateness and urgency which requires special attention. The MRP II system is helpful in providing a framework of perhaps a week or month production plan to work within, but it is still shop supervision which provides the hour-by-hour and minute-by-minute control.

The manufacturing engineer defines in detail the process plan to make a part. The task of marshalling the correct materials, transport, jigs, fixtures, gauges, instructions, etc., as well as operators and machine tools, is usually left to a variety of people, any of whom may make mistakes or forget. The result can be a series of delays leading to lost production time and lower efficiency. It has been estimated that as much as 80 per cent of the time a part spends in the factory is avoidable waiting time.

The CIMPICS system provides a set of facilities which enable parts to be controlled on their route through the factory to minimize waiting time and eliminate hold-ups due to missing tools or any other reason and thus reduces lead times and costly work-in-progress. CIMPICS allows the definition of the production process to be captured by the system. It can then marshal all the resources necessary at the lathe or milling machine prior to set-up. Requests and instructions are sent to the tool stores, transport and providers of paperwork in good time, automatically, and every time, for every batch.

The CIMPICS system has been installed in the small turbine disc cell to schedule and control production. The small turbine disc cell manufactures thirty-three different parts for various engines, including Pegasus, RB211, Tay and EJ200. The cell consists of thirty machine tools, fourteen of which are CNC/DNC controlled with some conventional millers, drillers, borers and grinders. The shop works a three-shift production cycle 24 hours a day, $6\frac{1}{2}$ days a week.

The system had to comply with Rolls-Royce's open systems philosophy, but manufacturing could not afford to use unproven technology. The result is a migratable open systems solution that delivers the required productivity savings.

7.9   System Infrastructure

The implementation of the CIMPICS system at Rolls-Royce Sunderland has been done in phases. The first phase installed the following.

(1) *Area controller*. An area controller receives instructions from a plant control system and schedules work for each manufacturing cell. Work schedules are then passed to each cell controller. The area controller holds a relational database used by the schedule compiler to check and marshal resources for each cell controller.

(2) *Cell controller*. A cell controller is the major element of a CIMPICS system which resides on or in the shop-floor area and exercises real-time control over the manufacturing environment. The cell controller receives the schedule or 'work-to-list' from the area controller, and then specialized software

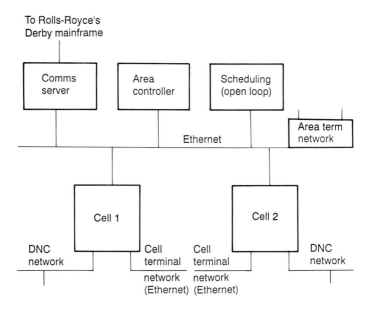

Figure 5.9   *Phase I*

developed using the CIMPICS Grafstar toolkit accepts and analyses the schedule for control and management of the shop-floor resources managed by the cell controller. The cell controller system can be hosted by any computer system that uses UNIX or UNIX-like operating systems with real-time extensions or facilities.

7.9.1   Broadband Infrastructure. A communications infrastructure broadband network was established around the factory. This provides the capability for many types of connections within the site and back to the Rolls-Royce Head Office in Derby via dedicated landlines. The broadband currently supports Ethernet to serve various networks of terminals, CAD/CAM systems, printers and computers in various departments and workshops. In the future it could also provide the infrastructure for MAP/TOP and indeed video security data.

7.10   Phase I Installation

The Phase I CIMPICS installation in Sunderland, which was operational until September 1990, provided the following services to the factory, as is summarized in Figure 5.9.

7.10.1   Scheduling. The mainframe data relating to works orders at Sunderland is passed to the area controller and its relational database via the comms server. The scheduling computer takes a snapshot of the most recent works order (MRP) requirement and then produces a feasible finite capacity

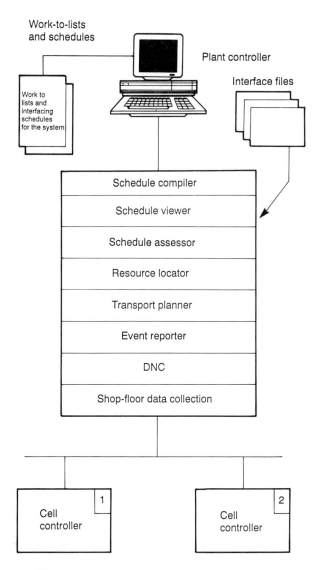

Figure 5.10 *Phase II*

schedule with the available resources of machine tools, operators, tooling
and material (work-in-progress).

In Phase I this was running a 'daily open-loop' mode with data relating to
works order completions being fed back directly to the Derby mainframe.
The scheduler then obtained the next coherent picture after the next nightly
update of works order requirements.

In Phase II the scheduler is running in 'closed-loop' mode with data being
fed to the area controller via the cell and its shop-floor date collection
terminals (see Figures 5.10 and 5.11).

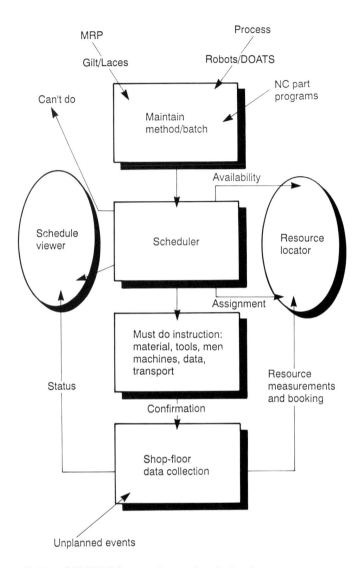

Figure 5.11   *CIMPICS overview – Sunderland*

7.10.2   Mechanized work booking (MWB). An existing shop-floor data collection system known as MWB is implemented at many Rolls-Royce sites. The CIMPICS MWB software provides a network of terminals on the shop floor supported by the cells. The data from operator input is then fed back to the MWB mainframe via the comms server.

7.10.3   DNC. The DNC system provides part program data to a variety of CNC machine tools, mainly vertical lathes, on the shop floor. Two

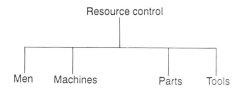

Figure 5.11 *CIMPICS overview – Sunderland*

networks, one from each cell, support 22 CNC connections. The network uses the DNet interface from the German company Dlog which was installed and commissioned by R.H.Symonds of Abingdon.

7.11 Phase II Implementation

The implementation of CIMPICS at Rolls-Royce Sunderland is a phased development, the Sunderland site always having been at the forefront of new systems designed to address the problems of advancing engine technology. Rolls-Royce Sunderland comprises four integrated factories covering more than 5 acres and is, in essence, a machining facility. Consequently machine tools are the key to its operating capability.

Both MERLIN and CIMPICS are concerned with logistics, but Sunderland needs technical instructions and a totally integrated manufacturing engineering system (TIMES) was launched giving manufacturing engineers online generation of planning documentation. These large, powerful systems supplement the growth in use of 'end-user' PC equipment and office link systems on the site.

Within Phase II, hardware configuration is being carried out, as it was found to be more efficient and cost-effective to maintain consistency of hardware. It will be a totally UNIX-based platform for the add-on modules supplied by Reflex, such as tooling, transport, shop-floor data collection, etc.

7.12 The Controlled Resources

Within any manufacturing environment there exists a number of different types of resources required to manufacture goods. These resources are divided within CIMPICS into the classifications of men, machines, parts and tools as shown in Figure 5.12.

Men are the human resource for a job and would normally be in charge of a machine, group of machines or processes. Machines are the physical devices which process the part being manufactured. Parts are the items which flow through the manufacturing process from machine to machine or process to process. Tools are the items required to set up and/or use in the manufacturing process and can be physical or electronic, that is a CAD drawing on a screen.

Physical tools such as cutters, gauges, fixtures, etc. are required for the setting up or the running of a manufacturing operation and are often shared between machines. They are stored at specific locations within the area or are 'in transit'.

The CIMPICS tool control system is designed to be used in both standalone and fully integrated modes. In the standalone mode, kit lists are generated manually and the user issues tools as he or she wishes. As there is no interaction with the CIMPICS schedule compiler, there is no guarantee that a tool will be available when required. In the fully integrated mode, requests for kit building are generated automatically by the resource manager task within the area controller.

The requirement for tools will previously have been calculated by the schedule compiler. The tools are then available for any task according to the information supplied to the compiler about tool availability in the short and long term, minimizing delays and 'lost' tools.

7.13   Phase II Improvements

The Phase II system improves upon the original 'Beta' version in a number of significant ways.

7.13.1   The computers. The computers have been upgraded to the latest RISC–open system UNIX machines with a degree of resilience and standby capacity. The lessons learned from the first phase – the need for a reliable, robust and high-availability system to support 24 hour production – have been learned and a number of small standard computers have been specified in preference to a larger centralized system. The advantages are the ability to adopt a 'black-box' strategy to implementation and the price performance edge of these almost commodity products.

7.13.2   The software. Resilience has been provided by dual disking, recording certain vital transactions, and the use of mirror-disc enabling software for the area controller database. A watchdog facility is provided to monitor the health of each system component, including the battery-backed power supply.

7.13.3   The network. The broadband and open system Ethernet networking philosophy has been maintained and extended to support additional terminals, printers and connection to the pre-existing CAD/CAM system via a network bridge.

7.13.4   Scheduling. The system is now 'closed loop' with shop-floor feedback available for immediate display and decision making at the area controller level.

7.13.5 Cell controllers. The original, somewhat proprietary, cell controllers with the MTOS real-time operating system have been replaced by 'standard' open system UNIX computers.

7.13.6 Shop-floor data collection. The software now prompts the operator to confirm the next set of transactions in a 'proactive' manner. That is, it identifies the next component or batch to be set up or manufactured and requests the operator to confirm that tools, materials, manufacturing instructions and CNC data are complete and available.

All the above-mentioned resources are marshalled prior to start by the area stores control and the transport control.

7.13.7 Transport control. This facility is planned for the later part of Phase II and instructs the transport personnel of the various collections and deliveries necessary from tool and material (WIP) stores to the machine tool workstations.

7.13.8 Tool stores. The tooling area is fitted with kitting instructions in advance of any set-up or manufacturing operation by the area controller during the marshalling phase.

7.13.9 DNC. The network has been extended somewhat to serve both the Small and Large Turbine Disc manufacturing areas.
'Paper tapes' have been eliminated from the shop floor and, as a consequence, the high availability of the DNC network is a prerequisite.
Two UNIX DNC servers are configured, each serving half the CNC connections. Each server, however, holds a duplicate of the data of the other and they are thus interchangeable. In the event of failure of one server, the two networks can be connected physically to the remaining DNC server and the service to the machine tools can be maintained, albeit at a slightly reduced response time in heavy-use situations.
Each DNC server is a 'standard open system' UNIX computer with the DNC network and connected via the computer VME backplane. The DNC to CNC interfaces, as before, are supplied by Symonds and Dlog.

7.13.10 MWB. The mechanized work booking (MWB) system is still supported with all transactions being fed back to the Derby mainframe. Those transactions logged by the CIMPICS cell control are converted to MWB format and then passed up via the comms server, as before.
The MWB service has been extended to cover both the Large and Small Turbine Disc cells and supports both operator and error investigation users.

7.14   Summary

Rolls-Royce, a world-leading aerospace manufacturer, has an ongoing strategy of investing in engineering and computer system techniques and infrastructure. The combination of sound production engineering methodologies, collaboration with other leading companies and the advantages to be gained from utilizing open system 'commodity' UNIX computers and networking, together with advanced software, are demonstrated in the Rolls-Royce Sunderland site.

The DTI-sponsored CIMPICS project, which installed a 'Beta' system in Sunderland, provided excellent experience and feedback and allowed a second, more comprehensive system to be installed which fits the user requirements for functionality and resilience.

The manufacturing systems engineering (MSE) philosophy implemented at Sunderland has focused the production objectives on substantially reducing work-in-progress and lead time and increasing stock-returns. The shop-floor control software has supported the cell-based reorganization and allowed a more positive approach to scheduling below the MRP level and operator data collection.

The next-generation MRP system being developed in house by Rolls-Royce, together with the close integration of CIMPICS shop control and the high priority of production engineering, is expected to reap even greater benefits in the coming years.

8   THE BENEFITS OF CIMPICS

The benefits which a system such as CIMPICS brings to manufacturing can be many and varied.

Uncoordinated manufacturing processes can leave work-in-progress items spending up to 85 per cent of shop-floor time waiting idly. Application of area/cell control can ensure that the short-cycle benefits of the JIT approach can be applied within an overall plan, thus cutting work-in-progress time to a minimum. This not only reduces order completion times and unit costs, but also gives a real competitive advantage from the lower lead times and more effective use of resources.

The CIMPICS environment and controllers are flexible and installation costs are low. Particular site implementation is achieved using straightforward, graphically presented menu screens, giving an immediate saving in implementation costs using company production engineers rather than external expensive software engineers. Subsequent modiciations can easily be incorporated. The use of international standards and unlimited external interfaces permits the user to adopt a unified system strategy. Each item of information need be entered only once, thus eliminating both wasteful duplication of effort and a major source of data transcription errors.

The area-wide real-time database provides the capability to obtain a true picture of resource capacity, and enables staff to manage data and produce accurate up-to-date reports.

CIMPICS can be reconfigured by the user and allows the system to grow with the organization. Individual cells can be planned, programmed and controlled, and the optimum or appropriate size of each cell can be determined entirely at the user's discretion. The internal system links and processes are defined by function, not by plant size or number of machines.

The Reflex CIMPICS system makes CIM affordable for companies who wish to implement manufacturing cells with few resources other than their own production engineers. CIMPICS makes CIM economically effective no matter what the scale of manufacturing operations: in small or medium-sized organizations for batch or series manufacturing, as well as individual cells or entire areas within much larger organizations.

REFERENCES

Cox B. J. (1986). *Object Oriented Language*. Reading, MA: Addison-Wesley.
IEC (1988). *'Preparation of Function Charts for Control Systems.'* IEC 848, OPEN.
Scheifler R. W. and Gettys J. (1987). The X Window system. *ACMJ., Trans. Graphics*, 5, no 2.

6 *The quick turnaround cell – an integrated manufacturing cell with process planning capability*

Tien-Chien Chang

1 INTRODUCTION

In the early 1970s, shortly after the first few flexible manufacturing systems (FMSs) were installed, it was predicated that thousands of FMSs would be installed by the mid 1980s. This optimistic prediction did not materialize. Now, almost 20 years later, only a few hundred FMSs have been installed. The reason is obviously due to the high initial cost and the inability of the existing systems to deliver what was promised by the FMS pioneers. These problems are caused by the complexity of the control and the scheduling system and the lack of other manufacturing supporting software. Since the conditions which inspired the development of FMSs still exist, and given the problems that current FMSs exhibit, a simpler and less expensive solution is needed. The manufacturing cell, which is less flexible and less automated, seems to be a viable alternative.

A manufacturing cell is a group of machines dedicated to produce one to a few families of parts. The material handling and machine control may be carried out manually. However, the individual machine in a cell may be automated. The hardware of a manufacturing cell is less complex than that of an FMS, and thus less expansive. The control and the scheduling of such a system is also simpler. However, the manufacturing supporting functions are still the same. In order to have a successful implementation of a flexible manufacturing cell (FMC) and gradually evolve it into a more automated system, the automation of the supporting functions is essential.

Manufacturing cells are generally designed for small batch production. In order to operate a manufacturing cell, a large number of process plans and part programmes need to be prepared. Process planning and part programming are just two of the many important manufacturing supporting functions. How to produce this information quickly is thus critical to the operation of the cell. A manufacturing system is used to realize a product. In any product realization process, excluding the business and the managerial functions, there are several operation functions: design, process planning, part programming, system control, execution, monitoring and final inspection. In order to run a manufacturing cell efficiently, it is necessary that the above-mentioned functions be integrated. Figure 6.1 depicts functions within a manufacturing system. The arrows in the figure show the

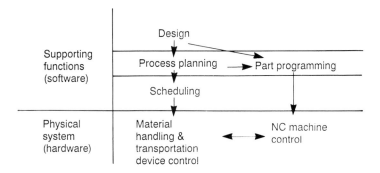

Figure 6.1   *Software and hardware functions in a manufacturing system*

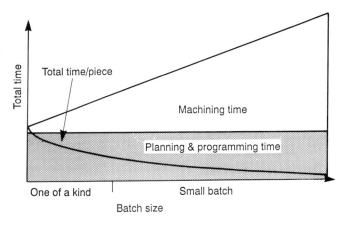

Figure 6.2   *Production time (planning, programming, and machining)*

information flow. Clearly, a manufacturing system cannot work without the supporting functions.

The following are some observations concerning FMC operation.

(1) In an FMC, although human operators are used for material handling and job scheduling, the planning and programming needs are the same as an FMS.

(2) The more flexible an FMC is, the more process plans and part programs need to be prepared.

(3) Unlike in mass production where process plans and part programs are readily available, in small batch production a major component of lead time is due to the delay in providing software. Figure 6.2 shows the total production time (without considering queuing delays) for a component. It can be seen that the effect of planning and programming time is especially critical to very small batch production.

A few general guidelines can be applied to the design of an FMC.

(1) Use general-purpose machine tools instead of specific machine tools.
(2) The more flexible the system is, the more responsive the system needs to be to changes.
(3) For a more flexible system, not only the machine tool control needs to be automated, but, more importantly, the planning functions need to be automated.
(4) An FMC should be designed for a specific domain of parts. When the domain of parts becomes too large, the tooling, fixturing, and planning problems become intractable.

2   SYSTEM REQUIREMENTS

A system called quick turnaround cell (QTC) has been developed at Purdue University to address the issues discussed above (Chang *et al.*, 1988). QTC not only consists of manufacturing cell hardware, but also includes the design, process planning, and inspection capabilities. The system is designed for a research and development and maintenance environment where the quick part turnaround is more important than the low production cost. The cost factor is considered only under the premise of rapid part production. The problem domain is limited to one-of-a-kind prismatic part machining, which is most challenging as far as process planning and inspection are concerned. The philosophy of a QTC is to respond to the demand quickly. The stochastic element of human-based planning in manufacturing is to be reduced to a minimum.

In order to achieve the objectives set forth, the following system requirements are determined.

- Integrated design/process planning/inspection system.
- Utilize the available resources to produce the part quickly.
- Other than design, no human decision making.

The above-mentioned requirements are primary requirements. Based on and derived from these primary requirements, there are also secondary requirements. They can be as important as the primary requirements for the successful implementation of the system. The following is a list of the secondary requirements:

- user-friendly design environment;
- no delay on stock preparation;
- minimum tooling preparation;
- detailed process plan generation;
- automated part program generation;
- adapting to desired changes in the cell capability;

- effective cell control and monitoring;
- automated inspection.

A user-friendly design environment is necessary in order to simplify the design. This design environment should interface with a solid modeller, so a complete and unambiguous design model can be generated for process planning use. Technological information such as tolerances and surface properties should also be part of the model and are easily entered through the design system. In a real production environment stock preparation can be a lengthy process. For the quick turnaround purpose, the delay on stock preparation should also be reduced to a minimum. The tooling preparation is the same as the stock preparation; the potential delay should be eliminated. A detailed process plan and finished part program can further reduce the possible human delay. Therefore, the automatic generation of such information is necessary. Further, the plans generated are to be executed quickly and should take advantage of the current cell status and resources. The machining cell must be properly controlled by the system, and its operation should be monitored. Finally, in order to ensure that the parts produced satisfy the design specification, they should be inspected quickly. To inspect one-of-a-kind parts quickly, it is necessary to automate this process.

3   SYSTEM INTEGRATION ISSUES

To integrate the design and planning functions means to share the same data among design and planning functions. Unfortunately, the design data is typically represented in a way which is easy only for the design analysis subsystems to use. In today's automated process planning system the design data cannot be used directly, and therefore a human must translate the data from the design model into a format suitable for process planning. The process plan generated by the system is not always easy for the system controller to interpret and use. This is but one example of the many such information roadblocks. At each stage of the product realization process, a data representation which is convenient for the isolated application is adopted. Before system integration can be achieved, it is essential to establish a high-level common view of the product data acceptable to all functions.

People have been trying for a long time to solve the vertical integration problem to link the design, process planning, device programming, and device control. Most efforts have overlooked the importance of the automation of information transfer. CAD, DNC, FMS, and CMM are some of the familiar solutions used; however, they have not been integrated into a system capable of operating without considerable human interpretation of the information. This human interface creates a bottleneck to the information flow. In a product realization process, process planning and device programming are two of the major information transfer tasks. In these two tasks, design information is transformed into manufacturing operation instructions (a process plan) and machine control data (part programs and robot programs). Only in recent years have automated process planning and part programming been studied extensively (Halevi et al., 1980; Chang and Wysk, 1985; Wysk et al., 1985; Kanumury and Chang, 1987; Alting and Zhang, 1989).

Process planning, as defined by Chang and Wysk (1985), is the act of preparing detailed operation instructions to transform an engineering design into a final part. To generalize the definition of process planning, part inspection planning can also be included. These both take CAD data as input and generate operation plans from these specifications. Currently there are two major schools of thought on what kind of CAD data to use as the input. One approach is to take a general CAD model (in order to provide complete and unambiguous data, often solid models are used) and develop an interface to recognize the manufacturing features from this model. This approach has been taken by Choi et al. (1984), Henderson (1984) and Joshi and Chang (1988). The advantage here is that a general modeller can be used for design. However, recognition is a non-trivial problem, as it is extremely hard to recognize complex features.

The second approach uses a specially designed CAD model incorporating shapes that are immediately recognizable by the manufacturing planner. These familiar shapes, or features, are designed around manufacturing operations. Such a *feature-based design* environment limits the designer to the available manufacturing features, which are by definition feasible for manufacturing. By using a feature model, one can ensure that parts designed are manufacturable. Macros for each feature can be written to synthesize operation plans and part programs. This approach has been taken by Eversheim et al. (Autap) (1980), Wysk (APPAS (1977), Chang and Wysk (CADCAM and TIPPS) (1985), Descotte and Latombe (GARI) (1981), Nau and Chang (SIPP, SIPS) (1985, 1986), Nau and Gray (1986), Hummel and Brooks (XCUT) (1986), Kramer and Jun (1986), Unger and Ray (1988), Cutkosky et al. (FirstCut) (1988), Hayes and Wright (IMW) (1989). The major shortcoming of the approach is the fact that the manufacturing planning job is moved to the design stage. Another limitation of the approach is that features used in the design must be functional shapes for the design task, instead of manufacturing shapes.

In QTC, a unique combination of both feature extraction and manufacturing-based feature modelling approaches was employed. A two-layered model is used: the higher-layer model carries more semantic information and the lower layer geometric information. Complete manufacturing information can be obtained by evaluating the lower layer with reference to the higher layer.

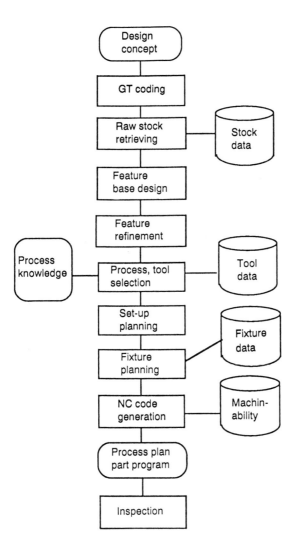

Figure 6.3 *QTC operation flow diagram*

4   SYSTEM APPROACH

In order to achieve the above requirements, the QTC research group proposed the following approach. Figure 6.3 shows the operation sequence of the system concept. To reduce the workpiece preparation time, precut raw stock is used. The solid model of precut stock is readily available in the design system. A designer selects the raw stock based on the approximate final part size. The design task is to select design features and size them and place them on the raw stock. The designer is working in the feature domain using boolean operators. The feature model, as it is called, is evaluated by a solid modeller and translated into a boundary model (boundary representation, or B-rep, is one common form of solid model) for display purposes. The feature model is what is passed along to the downstream manufacturing functions. Since the design features are not necessarily the final manufacturing features (an explanation is given in the following section), the feature refinement method is used to obtain the final manufacturing features.

Refined feature information is passed to an intelligent process planning system for process planning. The process planning system utilizes the cell status, such as available tools, fixture elements, etc. to come up with a final process plan; the cell status is maintained by a human operator. The process planning system also generates part programs and fixturing plans. The process plan is output in a standard process plan documentation format and part programs are written in a cutter location (CL) data format. The process plan also contains detailed information for the operation of the cell. The detailed set-up information can be displayed pictorially by the cell controller, thus providing guidance to the operator. The finished part is then inspected by a vision system.

The major advantage of this approach is the elimination of the delay due to human planning and programming. A less experienced machinist can be used to operate the cell since, regardless of who is doing the design or operating the machine, the final result is always consistent.

5   THE QTC SYSTEM ARCHITECTURE

The QTC system consists of four major functional modules: design, process planning, cell control, and inspection. Data files are used as the interface between these functional parts. This arrangement allows each functional part to be developed incrementally, without having a drastic effect on the overall system structure. There are two major human interfaces within the system: the first one is naturally the interface with the designer (process planning is done automatically, with no human interface necessary), and the cell control module has an interactive graphical user interface to facilitate detailed machine operator instructions. In the following sections, each of the four functional modules is discussed.

5.1   Design

As mentioned, the design environment is feature based, where each feature is loosely related to machining operations such as a hole, slot, or countersink (Turner and Anderson, 1988). The features currently implemented represent only material removal operations (depression features). Although protrusion features can be implemented in the design system as easily as depression features, they pose a very different problem to process planning and inspection. These problems will be discussed later.

One of the most important considerations is the design of the data view. How to make the design data easily usable by all system functions is of main concern. It was decided that a two-layered model should be used where the higher layer is the feature model just mentioned and the lower layer is a boundary model which is evaluated from the feature model by a solid modeller. The feature model is easy for the designer to use and the boundary model provides local information for process planning and the inspection system.

The objective of the design system is to provide an easy way to enter the feature-based model of a part. The feature model must contain complete information for manufacturing and a user-friendly environment must be provided.

The design module and design model data have been implemented using an object-oriented approach for dealing with features and their components. While all features have common physical attributes, such as position or orientation, they also have common actions they must perform. For example, they must be drawn or have their parameters changed interactively. By using an object-oriented approach, different features can be dealt with similarly. This isolates the 'local' characteristics of the individual features from the 'general' functioning of the system and facilitates the addition of new features.

In order to provide enough information for process planning to make the part, tolerance information is included in the feature model. Each feature has (unilateral) tolerances associated with its size parameters, such as depth or radius, and tolerances on its relative position. Also, there are form tolerances on features, such as surface finish or cylindricity. The tolerance scheme is represented by a position vector V_i and a tolerance matrix T_i:

$$V_i = \begin{bmatrix} DX_i \\ DY_i \\ DZ_i \end{bmatrix} \qquad T_i = \begin{bmatrix} TX_i^1 & TX_i^2 \\ TY_i^1 & TY_i^2 \\ TZ_i^1 & TZ_i^2 \end{bmatrix}$$

where DX_i, DY_i, DZ_i represent magnitude, TX_i^1, TY_i^1, TZ_i^1 negative tolerance and TX_i^2, TY_i^2, TZ_i^2 positive tolerance.

When there is a chain of tolerance (Figure 6.4), the position vector V_3 which defines the position of the last feature with respect to the reference is

$$V_3 = V_1 + V_2$$

and the tolerance is

$$T_3 = T_1 + T_2$$

However, if V_3 and T_3 are known, to find V_2 and T_2 one would use

$$V_2 = V_3 - V_1$$
$$T_2 = T_3 + T_1$$

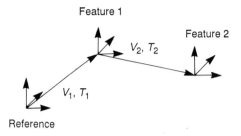

Figure 6.4   *Position and tolerance chain*

The tolerance is always additive. By applying this stacking method, a relative position can be found. Frequently, one feature is defined with reference to a different 'handle' on another feature. A position vector can be established between the two handles on the same feature, and then this position vector is treated in the same way as the position vector between two different features.

To assist the designer in interacting with the three-dimensional geometry of the features and in the construction of the model, graphical entities called handles are used. Handles are characteristic geometric elements of features, representing points and lines of interest. For example, point handles are used to represent the vertices of a rectangular workpiece or the end points of the axis of a hole, while line handles are used to represent geometric parameters, such as the length of a slot or the depth of a pocket. By interacting with a feature through its handles, the designer can easily position or orient the feature, or change the values of its geometric parameters.

Point handles represent geometric locations of interest on the features and can be used by the designer to establish reference points on these features. Corresponding to the way machining is done face by face, a default coordinate system is established on each face of the raw material where machining operations are possible. New references can be established by the designer on any point handle of a feature on the current work face or the workpiece which is also considered a feature. Other features can be positioned relative to this new reference, which establishes a positional relationship between the two features which is maintained in the feature model database. The absolute position of the new feature and its associated tolerance stack-up can be calculated from the relative position of the new feature and the absolute position of the reference feature. The final representation of a part is a data structure which consists of a list of features. This data structure is written to an ASCII file for storage. The part file is used to re-create the part data structure for use by the process planning, the cell control, and the vision inspection functions.

The design system calls the TWIN solid modeller (Marshburn, 1987) to produce a polygonal boundary representation of the part. During the design

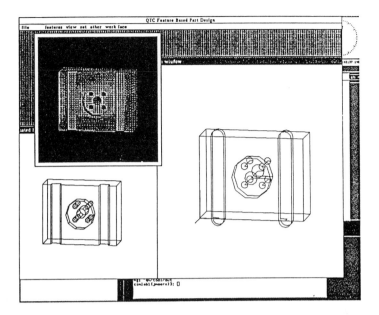

Figure 6.5   *Design window display*

stage, TWIN is used to generate a boundary representation for display and verification for the designer. Two graphics packages display the picture in either wireframe or shaded image of the evaluated boundary representation. The only data passed on the subsequent system functions is the feature model consisting of a list of features, starting with the workpiece, their parameters, and the handle-based positioning information. The boundary model is regenerated as it is needed.

Figure 6.5 shows a screen dump of the design windows. On the right-hand side is the feature model, and each designed feature is shown. A wireframe drawing of the B-rep is shown on the lower left corner whilst the upper left corner shows a shaded image of the final part.

No formal engineering drawing is generated by the system at this time. Actually, there is no need to provide the three-view drawing.

5.2   Process Planning

The objective of process planning is to be automatic and accurate. A system called AMPS (Automatic Machining Planning System) (Kanumury, 1988; Kanumury *et al.*, 1988) has been developed for process planning and consists of four functional modules: feature refinement, process selection, fixturing method planning, and NC cutter path generation. The first module does feature refinement, working from the design model. The process planning system must know each machining feature's exact shape and size, its possible tool approach and feed directions, and the relationship between

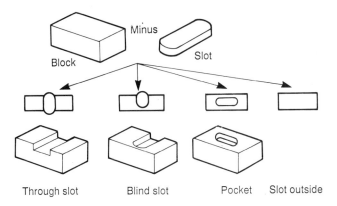

Figure 6.6 *Feature reclassification*

pair of such features. For example, consider a simple case where a part is represented by a straight slot with some attributes (Figure 6.6). Depending on the location and the size of the slot, it can be considered an open slot, a blind slot, or a pocket. It is necessary to find out the above information before one starts the subsequent process planning tasks.

To answer these questions, information about the face must be known. Since the design is given as a feature model which contains a list of features, their parameters and attributes, it is necessary to evaluate it into a boundary model. The feature refinement problem is somewhat similar to that of the feature recognition problem. It is especially true when dealing with protrusion features and, in this case, the machined feature is the volume surrounding the designed protrusion feature. The TWIN solid modeller provides a custom pointer for every geometric entity. A tag (data item) is attached to each face of the feature by using the pointer, and tags are inherited by the evaluated boundary model. This tagging thus makes it simple to trace back from a face in the boundary model to the feature which is its origin, making the recognition of final feature shapes easier. Since the feature class is explicitly defined by the designer, the recognition is a subclassification. Rules have been written to place the designed features into their subclasses.

In addition to identifying the final feature type and feature boundary, the feature refinement module also finds feature interactions such as 'contained-in' and intersection-type information. Both of these types of information are used as constraints for process sequencing. The output of the feature refinement module is refined features and precedence relations represented symbolically. This data is then passed to the second module (process selection) and the part data at this time is represented by frames. An example of a feature frame is shown below.

The feature is 'slot_1' which is of type '1' (straight slot). In the next section, the b_type slot indicates that the bottom of the slot is flat. Reference feature (ref_feat) and reference face (ref_face) slots point to the feature in which the current feature is referenced. Following this section are surface finish and tolerance values. Not all tolerances have been defined, for instance angularity. Approach directions are given in the slots following and each direction is defined by a vector. In the feed direction slot, there are two vectors showing two feed directions (feature relations are given in the following section). This feature contains at its bottom a feature, 'hole1', the current feature is contained by feature 'slot2', and there is no merging nor splitting of the feature.

```
(slot_1
        ((id 2)
        (type 1)
        (feat_type 0)
        (length 4.50000)
        (width 0.80000)
        (depth 0.500000)
        (b_type FLAT)
        (handle_face_ref 0)

        (ref_feat work_piece)
        (ref_face face1)

        (surface_finish 250.000000)
        (length_tol_range (-0.050000 0.050000))
        (width_tol_range (-0.050000 0.050000))
        (depth_tol_range (-0.050000 0.050000))

        (locational_tol_range (-0.086603 0.086603))
        (parallelity_tol_ref ())
        (perpendicularity_tol_ref ())
        (angularity_tol_ref ())

        (approach_direction ((0.000 0.000 -1.000)))
        (aux_approach_direction ())
        (feed_direction ((0.000 1.000 0.000)
                (0.000 -1.000 0.000)))
        (thru_direction ((0.000 1.000 0.000)))
        (thru_type THRU_ONE_END)

        (contains_b (hole1))
        (contains_s ())
        (contained_b (slot2))
        (contained_s ())

        (merge ())
        (split ())
        (causing_split ())))
```

The process selection and fixturing method planning are done by an expert system. A backward planning approach is used. Backward planning means that the planning process begins at a finished part and tries to find a set of processes to fill all the features. Process capabilities rules are modelled by *if . . . then* statements. The antecedent of a rule is what the process can accomplish under the best operating condition, while the consequent shows what minimum condition the feature surface must be in before a process can be applied. For example, a boring process requires a smaller hole than the desired hole before the process can be used. The antecedent of the boring process is the final hole diameter and the consequent has the final hole diameter minus the clearance. When the process is selected, it will fill the hole with a layer of material. Both forward and backward chainings are used in searching for the goal. The chaining direction depends on the specific subgoal being sought.

Process capabilities based on previous research results, and input from industry, have been collected as process knowledge which is organized in a hierarchical fashion. A process taxonomy classifies the processes into hole-making and non-hole-making processes. The hole-making processes are further classified into fine-finishing operations, finishing operations, hole-generation operations etc. The hole-generation operations includes conventional processes, deep-hole processes, etc., and the conventional processes include centre-drilling, core drilling, twist drilling, end milling. etc. The antecedents of the higher-level classes are inherited by the lower level. The major advantages of this approach are that it is easy to understand and it reduces the search space. The process knowledge is also better organized for maintenance. For example, a rule for a rough boring process is shown below written in the KEE language (KEE stands for 'Knowledge Engineering Environment', an expert system toolkit from Intellicorp) embedded with LISP functions. In this rule the variable ?FEAT refers to a frame representing the feature being planned. The process is first described as a type of FINISHING_OPERATIONS process. The process hierarchy is defined by the CLASS. In the antecedent of the rule, first the feature type is checked and then a LISP function is executed to check the technological constraints of the process. In the consequent, first a new frame is created by IN.NEW.WORLD then LISP functions are used to change the hole radius.

```
(ROUGH_BORING
     (FINISHING_OPERATIONS (CLASSES GENERICUNITS))
     NIL
     ()
     ((ACTION.TYPE (TOGETHER))
      (EXTERNAL.FORM
     ((IF   (TEXT (PROCESS PROGRESS))
            (?FEAT IS IN CLASS PRIMARY_HOLE_FEAT)
            (THE L_RADIUS OF ?FEAT IS ?LRA)
            (LISP (ROUGH_BORING_SATISFY_CONSTRAINTS
                  ?FEAT))
            THEN
            IN.NEW.WORLD
            (LISP (ROUGH_BORING_MODIFY_RADIUS
                  (- ?LRA 0.08)
                      0.1
                      ?FEAT
                      $WORLD$))
```

```
        (LISP (ROUGH_BORING_MODIFY_CONSTRAINTS
            ?FEAT $WORLD$))
        (THE PROCESS OF ?FEAT IS ROUGH_BORING)
        (LISP (QUERY (PROCESS.GOAL ?FEAT)
                'HOLE_MAKING_RULES
            $WORLD$) ) ) ) )
    (MAKE.AND.WORLD? (NIL) )
    (PARSE.ERRORS)
    (RULE.TYPE (NEW.WORLD.ACTION)
    (WEIGHT (5.0) )
    ) )
```

The process selection module selects a set of feasible processes for each feature. Initially depth-first search is used. Since rules are arranged in the order of preference (based on cost and availability), the final result is a reasonable set of processes. The most appropriate cutting tools are then selected. Tool selection rules are used to synthesize tool database queries, and returned by each of these queries is a set of feasible tools. The tool database, like raw stock, fixture, and machinability databases, uses a relational schema. In each data record information is stored on tool number, tool material, tool geometry, dimension, tool life left, and operation types.

Tools are also classified into three groups: those which are currently in the tool magazine, those which have already been set up, and those which are available in the tool crib but have not been set up yet. When tools are selected the priorities are given according to the sequence mentioned above. The rationale is that minimum tool preparation time should be spent in order to minimize the lead time for the part manufacturing. If an exact tool cannot be found, a tool second on the list is selected. The system also tries to use the same cutter for several features. The selection can be controlled by providing the system only a smaller set of tools in its database. Although the one cutter may not be the best for all of the features, reducing the number of tools needed for the job reduces the number of tool changes. In machine shops, if only optimum cutters are used, hundreds of cutters need to be purchased and maintained. This represents a large amount of capital investment and tremendous set-up and storage management problems. Also the tool magazine can hold only a limited number of tools at any given time, and the loading and unloading of additional tools creates undesirable idle time.

The next step is to determine the set-ups required and the features to be machined under each set-up. A feasible set-up position is determined by the maximum number of uncut and available features which can be approached by the cutters. The availability of features is determined by the precedence constraints. Figure 6.7 shows the set-up determination approach.

A fixturing method planner (the third module of AMPS) is used to determine how to hold the workpiece for the position and orientation specified by the set-up planner. It must be kept in mind that the shape of the workpiece changes as more set-ups are done. The original fixturing method

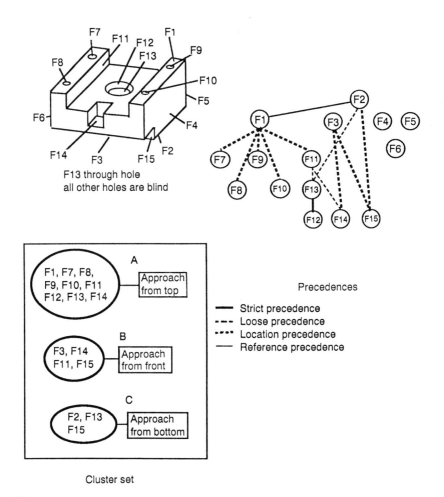

Figure 6.7 *Set-up determination*

planner is call CLAMPS (Shah, 1988). The purpose of CLAMPS is to generate a fixturing plan which ensures a firm holding of the workpiece and a minimum number of refixturings. The fixturing position also allows the maximum number of features under a set-up to be machined. The fixturing method planner is incorporated into the process planning system and interacts with the set-up planner. There are three major parts in a fixturing planning system: selection of locating and clamping points, stability analysis, and fixture configuration. A generation, test, and rectification procedure has been developed in which locating and clamping points are selected using some heuristics, and the result analysed. If the result fails, the points are modified. Finally, fixture elements are selected to satisfy the proposed locating and clamping configuration.

In the system, fixtures are classified into vice-jaw fixtures, angle-plate fixtures, plate fixtures, indexing fixtures, etc. Each class is decomposed into basic fixture functional elements, and each element is then classified further. Rules are written to relate the functional elements to the part geometries and the required work-holding method. Currently only vice-jaw-type fixtures have been implemented and tested, but a modular feature configuration system is being implemented. The final fixturing plan is shown graphically by the system. Figure 6.8 shows a display of a fixturing plan where, for each set-up the system can show the workpiece before and after machining. The pictures can be used to guide the operator on loading the workpiece into the fixture, or, in the case of a modular fixture, to assemble the fixture.

At this stage set-ups, fixturing method, processes, tools, and preferred machining sequence have been determined. The next step is to finalize the machining sequence and to generate NC part programs automatically. It is necessary to generate a part program for each set-up. Algorithms have been developed to generate the cutter path while considering minimizing cutter motion. Currently the minimum clearance height for a given set-up is detected and used for tool clearance under that set-up. The cutter path generation module uses the features faces, and the workpiece location and orientation. The probe on the machining centre is used to detect the exact location and orientation of the workpiece in the fixture, thus compensating for any error. A probing algorithm detects the maximum distance points on the probing surfaces, the latter being the front and the two side surfaces. The farther apart the probing points, the better the accuracy.

The final result of the process planning function is process plan documentation and several part programs written in a cutter location data format (Chang, 1987a,b). These data files are used by the cell control module. In order to modularize the system, a process plan documentation language was developed. Using this language, all information pertinent to the machining of a part is available, which allows for independent future extensions to the process planning and cell control modules, without suffering from format changes. An example output of the process planning module is shown in Figures 6.9 and 6.10. In Figure 6.9, it can be seen that each section of the process plan is separated by a keyword and begins with a '%' sign. For example, the machine tool information is denoted by '%%C2' followed by an explanation. Relevant data is presented in each section, for instance in the machine tool information section, the machine tool selected (keyword MCNA) is T_10_CNC and the id number (keyword MCNO) is 1211212112. The CL data in Figure 6.10 is numerically coded. The first digit in a line is the code. Code 1 stands for coordinates, code 2 is the feed rate, etc.

5.3   Cell Control

The purpose of the QTC cell controller (Moore and Chang, 1988) is to control and coordinate the cell devices, schedule the jobs, maintain

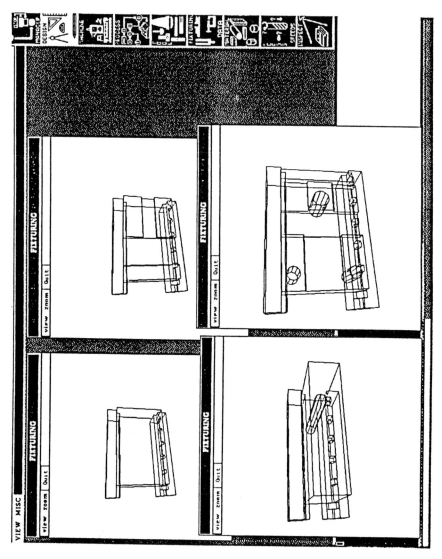

Figure 6.8 *Fixturing method display*

databases, and provide an interactive operator interface (Figure 6.11). A human operator is used in the cell for tool and part loading/unloading, and extensive windows, icons, and pull-down menus are used to guide the operator. For example, Figure 6.8 illustrates the fixturing plan display under the cell control window. On the right-hand side of the window are icons representing the cell controller functions, so minimum keyboard typing is required.

Whenever a process plan is produced by the planner, the cell controller

%A "WORKPIECE INFORMATION"

%%A1"PART"
PTNO 10 PTNA "DEMO"

%%A2"TECHNOLOGICAL"

MTDS "FMCS_WROUGHT_LC_RESULFURIZED"
WPTY P WBHN 0 WPSH P
WPLE 3.0 WPWI 5.0 WPHI 1.2

%C "PROCESS PLAN INFORMATION"

%%C1 "PRE-LOADING INFORMATION"
MTSQ 1
PMHD M NOSP 3

%%C2 "MACHINE TOOL INFORMATION"

MCNO 1211212112 MCNA "T_10_CNC"

%D "MACHINE TOOL LEVEL DESCRIPTION"

%%D1 " SETUP INFORMATION"
STNO 1001
STFP "SETUP_0"

%%D6 "WORK PIECE ORIENTATION"
WPOP 15.684 7.460 8.296
WPOR 1.5708 0.000 3.141

%%D9 "NUMBER OF OPERATIONS"
NOOP 3

%E "OPERATION LEVEL DESCRIPTION"

%%E1 "MACHINING SPECIFICATIONS"
OPNO 100101
OPDS PRB

%%E2 "OPERATION TOOL NUMBER"
OPTN 18

%%E4 "OPERATION DATA"
OPSR OFF OPCT OFF
OPFN "SETUP_0"

%E "OPERATION LEVEL DESCRIPTION"

%%E1 "MACHINING SPECIFICATIONS"
OPNO 100102
OPTY R OPDS MSL

%%E2 "OPERATION TOOL NUMBER"
OPTN 10

%%E4 "OPERATION DATA"
OPSP 104.7 OPDF 9.480469 OPDP 0.765
OPSR CLW OPCT FLE
OPFN "SETUP_0"

%E "OPERATION LEVEL DESCRIPTION"

%%E1 "MACHINING SPECIFICATIONS"
OPNO 100103
OPTY R OPDS MSL

%%E2 "OPERATION TOOL NUMBER"
OPTN 4

%%E4 "OPERATION DATA"
OPSP 104.1 OPDF 2.317128 OPDP 0.265
OPSR CLW OPCT FLE
OPFN "SETUP_0"

Figure 6.9   *Process plan documentation*

processes it immediately. The entire documentation is parsed, so the required resources (machine, raw material, tools, and fixtures) are identified. The appropriate post processor is used to convert the cutter location files into machine-specific part programs. Each document identifies a new job which is then placed in the machine queue automatically by the controller. Tools are provided for the operator to manipulate the queue. At any time, the controller can display the required resources, the part design, the features to be machined, and a drawing of either the workpiece or the finished part being clamped in a fixture.

5.4   Vision Inspection and Monitoring

An integrated vision system has been designed that uses the design-level feature database to plan and execute algorithms which attempt to determine whether the final part is within specifications.

```
15 0                                        1 11.1840  10.1064  8.6496
11 2      1                                 11 1
1 25.0250  6.1221   12.0000                 1 11.1840  8.1536   8.6496
6  1      18                                1 11.1840  8.1536   9.1424
1 25.0250  6.1221   12.0000                 16 1      1
11 1                                        11 3      1     1       14.6840
4 0                                         11 5      1     1     1.     13.6840
3 150.000   1                               1 11.1840  8.1536   9.1424
1 25.0250  10.1064  12.0000                 6 0       13
1 14.0376  10.1064  12.0000                 4 1
1 14.0376  10.1064  11.9960                 16 1      1
16 1      1                                 1 11.1840  8.1536   12.1424
11 3      1     3       8.4960              15 3      11.1840  8.1536  12.1424  180
11 5      1     3     1    9.4960           1 11.1840  8.1536   12.0000
1 14.0376  10.1064  11.9960                 6 1       13
11 1                                        1 25.0250  6.1221   12.0000
1 14.0376  7.8136   11.9960                 1 25.0250  8.9600   12.0000
1 18.3304  7.8136   11.9960                 1 17.1840  8.9600   12.0000
16 1      1                                 1 17.1840  8.9600   9.5960
11 3      1     3       8.4960              7 1       641
11 5      1     3     1    9.4960           3 9.539    1
1 18.3304  7.8136   11.9960                 5 1
6 0       13                                9 6       0
11 1                                        1 17.1840  8.9600   -1.6175
4 0                                         1 0.0000   0.0000   9.5960
3 150.000   1                               6 0       23
1 18.3304  10.1064  11.9960                 4 1
1 12.5840  10.1064  11.9960                 1 17.1840  8.9600   11.6960
1 11.1840  10.1064  11.9960                 1 17.1840  8.9600   12.0000
1 11.1840  10.1064  8.6496                  6 1       23
16 1      1                                 1 25.0250  6.1221   12.0000
11 3      1     1       14.6840             1 25.0250  10.0600  12.0000
11 5      1     1     1    13.6840          1 14.2840  10.0600  12.0000
                                            1 14.2840  10.0600  9.5960
                                            7 1       835
                                            3 9.473    1
```

Figure 6.10   *CL data file*

The hardware used for the system includes CCD video cameras and VME board-level image processing components supplied by Imaging Technology. The vision cards interface directly with the VME bus on the Sun workstation. The tool inspection and the finished part inspection use various image processing and computer vision algorithms. However, the most important component of the vision system is the integration of the ability to plan what and where to inspect based on the design model.

The finished part inspection system (Park and Mitchell, 1988) receives refined feature data from the process planning system and uses the TWIN solid modeller to generate the necessary boundary data structure for part inspection. Currently a two-dimensional vision system is used for part recognition and inspection which first prioritizes a set of visible design features (e.g. through hole, slot) based on their expected visibility (through direction as mentioned in feature refinement). The system selects an algorithm from a list of image processing feature detection algorithms. For example, if from the boundary represention only hole features are present in the part, only hole detection and corner detection algorithms are needed. These algorithms are applied to the image to detect the features. The

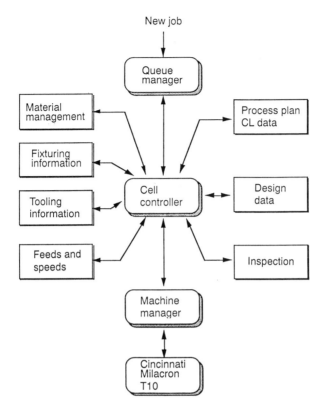

Figure 6.11   *Cell controller functions*

features in the image and those in the part model are matched using a depth-first tree search algorithm. This matching is made significantly easier by the fact that the B-rep edge, vertex, and face information has appended to it the symbolic high-level design feature information. Thus when image features are found to match a segment of the B-rep model, the corresponding design features which produced the image features can be readily identified. The matching result verifies the correctness of the part and also provides the information for the camera transformation matrix. Thus the part does not have to be precisely placed on the inspection table for accurate inspection to occur.

   After the image features are matched to the model, the final precision inspection is done. Precision measurement algorithms are applied to the regions of the image and the results are compared with the dimensioning and tolerance information stored with the design features. Precision measurements can be made for edges, perimeters, areas, circle centres, etc. (Mitchell *et al.*, 1986). The inspection planning system must choose the algorithms to be applied to determine whether the part meets the specifications. Since the present vision system is only two-dimensional, some depth measurements

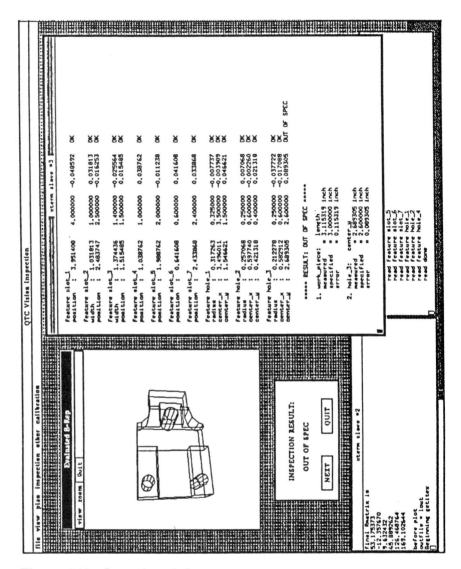

Figure 6.12   *Inspection window*

cannot be made. It is planned to add stereo cameras and a structured light source to the inspection system so that it can handle some of these measurements. Figure 6.12 shows the result of an inspection.

6   SYSTEM IMPLEMENTATION

Both hardware and software environments were determined early in the project. The interfaces between software modules were also defined.

Initially the Sun 3 workstation was chosen as the hardware platform because it is a UNIX-based machine and uses the same LAN (Local Area Network) protocol as the existing engineering computing network at Purdue. The X-Window windowing system and the CommonLips and C languages were adopted as the software implementation environment. This environment will allow the software to be ported on to many different computers in the future. The KEE expert system shell was used for the process selection and fixturing planning modules. KEE has provided a powerful knowledge-modelling capability, yet requires a large amount of memory and a different windowing system. For this reason, the first-generation QTC system requires at least two Sun workstations to run. In a new version, a drastic change in both the hardware and the system environment has been made. A Silicon-Graphics Personal IRIS workstation replaces the Sun workstations and KEE is replaced by CommonLips-based code. Although the replacement of KEE took a great deal of effort, the other changes have proven to be less painful. The main change is on graphics. On the new platform all the software is able to be run under the same environment. Currently the system consists of a large amount of computer codes written in C and LISP, and no commercial software package is used. With the limited design features used initially, the system has been able to produce perfect parts in minutes, without any human technical decisions required after the design is completed.

7   CONCLUSIONS

In this chapter an integrated design/process planning system – QTC – was introduced. The QTC system was designed with the goal of being used in a one-of-a-kind prismatic part machining environment. The system is able to generate a sound detailed process plan and NC cutter path automatically. What makes the system unique is the ability to generate a process sequence automatically and correctly. The system is by no means complete, but this general framework is able to accommodate new additions.

The QTC system approach is not without some problems, the major problem being its complexity. To make a good process plan, one needs a lot of information and experience. When implemented in a computer system, this information and experience needs to be programmed in. From the discussion in this chapter, one may see that a tremendous amount of data and knowledge has been used in this prototype system. Data can be collected much easier than knowledge, but to be useful, both knowledge and data have to be represented in the system. The major difficulty is the geometry-related knowledge has been used in this prototype system. Data can be collected much easier than knowledge, but to be useful, both knowledge and data have to be represented in the system. The major difficulty is the geometry-related knowledge, for geometric reasoning is hard and tedious. There is no unified approach for all geometric reasoning problems, and there may never be one.

Many of the problems have to be treated as special cases and therefore algorithms need to be developed for these special cases. It can be a very difficult task to develop all the necessary algorithms, especially when the domain of the features and processes attainable by the system increases. Before a generic geometric reasoning method is developed, a system like QTC is best used for parts which do not require very complex shapes. On closer investigation one may find a large number of parts which fit into this category.

Although the QTC system as it stands is not sophisticated enough to replace an experienced process planner/machinist, it can well be a junior member of a manufacturing team who handles the less complex parts. Experience and common sense suggest that the majority of parts produced in industry are not too complex and thus fall into the domain of QTC. Thus, the QTC system can work well with humans to relieve them of the majority of the workload and give them time to concentrate on more difficult and challenging problems.

The approach, which utilizes a common design model–feature model for design, process planning, and inspection, has been proven to be effective. The ability to reason about the final machined feature geometry and its relationship to other features enables the system to make better plans and at the same time relieves designers from considering excessive manufacturing details. With minor additions to the system, the exact machining time and cost will be predictable, thus providing designers with precise feedback. A simulation system will also provide realistic machining operation simulation in real time. Since the process plan documentation is in ASCII format and is keyword driven, a human planner can easily modify it. When the system is used for generating plans for mass-production parts, it can provide better feedback to the designers and allow humans to make changes to the results. The fast response time (compared with humans) and consistent performance enable the design/prototyping/testing/evaluation cycle to be shortened, thus giving a competitive edge.

It may be concluded that a system like QTC works best for medium complexity to simple parts. Its use is not limited only to quick turnaround environments but also to any other process planning environment.

ACKNOWLEDGEMENTS

The work described in this chapter was supported by the National Science Foundation Engineering Research Center on Intelligent Manufacturing Systems at Purdue University and its industrial partners and affiliates. The design module was developed by Professor D. C. Anderson and the inspection system by Professor O. R. Mitchell initially and currently by Professor A. Kak. Mr James Moore developed the original cell controller and is the machine operator. The QTC system would not be a reality without the diligent work of many students who were once QTC group members.

REFERENCES

Alting L. and Zhang H. (1989). Computer aided process planning: the state-of-the-art survey. *Int. J. Prod. Res.*, **27**, 553–85.

Chang T. C. (1987a). Integrated process planning approach for discrete part machining. *Proc. 14th NSF Conf. on Production Research and Technology*, Ann Arbor, Michigan, 6–9 October, pp. 43–50.

Chang T. C. (1987b). *MAPT 3.0 user's manual*. School of Industrial Engineering, Purdue University, W. Lafayette, IN.

Chang T. C. and Wysk R. A. (1985). *An Introduction to Automated Process Planning Systems*. Englewood Cliffs, NJ: Prentice-Hall.

Chang T. C., Anderson D. C. and Mitchell O. R. (1988). QTC – an integrated design/manufacturing/inspection system for prismatic parts. *Proc. ASME, 31 July–3 August 1988, Computers in Engineering Conf.*, San Francisco, CA, Vol. 1, pp. 417–26.

Choi B. K., Barash M. M. and Anderson D. C. (1984). Automatic recognition of machined surfaces from a 3D solid model. *CAD*, **16**, no 2, pp. 81–86.

Cutkosky M. R., Tenenbaum J. M. and Muller D. (1988). Features in process-based design. *Proc. 1988 ASME Int. Computers in Engineering Conf. and Exhibition, ASME, 31 July–4 August*, pp. 557–62.

Descotte Y. and Latombe J.-C. (1981). GARI – A problem solver that plans how to machine mechanical parts. *Proceedings of the 7th International Joint Conference on Artificial Intelligence*, August 1981, pp. 329–347.

Eversheim W., Fuchs H. and Zons K. H. (1980). Automatic process planning with regard to production by application of the system AUTAP for control problems. *Computer Graphics in Manufacturing Systems, 12th CIRP Int. Semin. on Manufacturing Systems, Belgrade*.

Halevi G. *et al.*, (1980). Development of flexible optimum process planning procedures. *Ann. CIRP*, pp. 313–17.

Hayes C. C. and Wright P. K. (1989). Setup planning in machining: an expert system approach. *Proc. 1989 NSF Conf. on Advances in Manufacturing System Integration and Process, Berkeley, CA, January 1989*, pp. 441–3.

Henderson M. R. (1984). Feature recognition in geometry modeling. *Proc. CAM-I's 13th Ann. Meet. and Technical Conf.*, 13–15 November 1984.

Hummel K. E. and Brooks S. L. (1986). Symbolic representation of manufacturing features for an automated process planning system. *Bound Volume, Symp. on Knowledge-Based Expert Systems for Manufacturing, ASME Winter Ann. Meet., Anaheim, CA*, 7–12 December 1986.

Ingrand F. and Latombe J.-C. (1984). Functional reasoning for automatic fixture design. *CAM-I Technical Conf.*, November 1984, vol. 8, pp. 53–65.

Iwata K. *et al.* (1980). Development of non-part family type computer aided production planning system CIMS/PRO. In *Advanced Manufacturing Technology (Proc. 4th Int. IFIP/IFAC Conf., Prolomat 79, 1979)*, (Blake P. ed.), North-Holland, New York: Elsevier.

Joshi S. and Chang T. C. (1988). Graph-based heuristics for recognition of machined features from a 3-D solid model. *Computer-Aided Design*, **20**, no. 2, pp. 58–66.

Joshi S., Vissa N. and Chang T. C. (1988). Expert process planning system with solid model interface. *Int. J. Prod. Res*, **26**, no 2, pp. 863–885.

Kanumury, M. (1988). AMPS – an automatic manufacturing process planning system. *MSc Thesis*, School of Industrial Engineering, Purdue University

Kanumury M. and Chang T. C. (1987). *Survey of Process Planning Sustems for Turned Parts*. Engineering Research Center on Intelligent Manufacturing Systems, Purdue University, January 1987.

Kanumury M., Shah J., and Chang T.C. (1988). An automatic process planning system for a quick turnaround cell – an integrated CAD and CAM system. *USA-Japan Symposium on Flexible Automation, ASME*, July 18th, 1988.

Kramer T. R. and Jun J.-S. (1986). *Software for an Automated Machining Workstation*. Report, National Bureau of Standards, July 1986.

Mashburn T. (1987). A polygonal solid modeling package. *MSc Thesis*, Purdue University.

Mitchell O. R. *et al.* (1986). Recent results in precision measurements of edges, angles, areas, and perimeters. *Proc. of SPIE on Automated Inspection and Measurement, vol. 730*, October, Cambridge, MA, pp. 123–34.

Moore J. and Chang T. C. (1988). The control of a quick turnaround cell – an integrated CAD and CAM system. *10th Conference for Computer and Industrial Engineering*, Dallas, TX, March 23–25, 1988, pp. 315–323.

Nau D. S. and Chang T. C. (1985). A knowledge based approach to process planning. *Bound Volume of the Symp. on Computer Aided/Intelligent Process Planning, The Winter Ann. Meet. of the ASME*, Miami Beach, Florida, 2–17 November 1985, pp. 65–72.

Nau D. S. and Chang T. C. (1986). Hierarchical representation of problem-solving knowledge in a frame-based process planning system. *J. Intell. Syst.*, **1**, 29–44.

Nau D. S. and Gray M. (1986). SIPS: an approach of hierarchical knowledge clustering to process planning. *Bound Volume, Symp. on Knowledge-Based Expert Systems for Manufacturing, ASME Winter Annu. Meet.*, Anaheim, CA, 7–12 December 1986.

Park H. D. and Mitchell O. R. (1988). CAD based planning and execution of inspection. *Proc. IEEE Computer Vision and Pattern Recognition Conf.*, Ann Arbor, MI, 5–9 June 1988.

Shah J. (1988). CLAMPS – automated fixturing in a flexible manufacturing environment. *MSc Thesis*, School of Industrial Engineering, Purdue University.

Turner, G. T. and Anderson D. C. (1988). An object-oriented approach to interactive, feature-based design for quick round manufacturing. *Proceedings the 1988 ASME Computers in Engineering Conference*, San Francisco, August 1–4, 1988, pp. 551–556.

Unger M. B. and Ray S. R. (1988). Feature-based process planning in the AMRF. *Proc. 1988 ASME Int. Computers in Engineering Conf. and Exhibition, ASME*, 31 July–4 August 1988, pp. 563–9.

Wysk R. A. (1977). An automated process planning and selection program: APPAS. *PhD Thesis*, Purdue University.

Wysk R. A., Chang T. C. and Ham I. (1985). Automated process planning systems – an overview of ten years of activity. *Proc. First CIRP Working Semin. on Computer Aided Process Planning*, Paris, France, 22–23 January 1985.

7 Non-hierarchical cell control

Neil A. Duffie

1 INTRODUCTION

Designers of automated manufacturing cells are faced with the task of integrating heterogeneous groups of machine tools, robots, guided vehicles and other production equipment into functionally homogeneous systems and subsystems. A large number of alternatives exist in the design of these systems and their interactions with the larger application environments in which they reside. There is a spectrum of possible manufacturing cell control architectures as shown in Figure 7.1. Current designs are hierarchical and are constructed using a philosophy of 'levels' of control. Entities at each level in the hierarchy make decisions based on commands received from the level above, and sensory feedback received from the level below. It has been noted by Hatvany and Nemes (1978) that for systems above a certain size and complexity, it becomes impractical to employ a deterministic system design approach in which 'precise, functional descriptions are established for each system and of all subsystem interconnections' together with the creation of 'a clear-cut hierarchical structure in which each and every one of the subsystems and the interconnections is unequivocally implemented'. They recognized that in large, complex systems it is not feasible for the designer to specify each and every subsystem interconnection and operational detail.

In a non-hierarchical architecture, relationships between entities in the system are not master/slave, and entities cooperate with each other without regard to 'level'. Non-hierarchical control architectures offer prospects of reduced complexity and improved fault tolerance by localizing information and control, reduced software development costs by eliminating supervisory levels, and higher maintainability and modifiability due to improved modularity and self-configurability (Clark and Svobodova, 1980; Duffie, 1982, 1987; Enslow, 1978; Hatvany, 1985; Kleinrock, 1985; Vamos, 1983). Unfortunately, traditional engineering methods used to develop and implement computer software and hardware for manufacturing control applications are not necessarily appropriate for the design of these systems. The non-hierarchical architecture therefore raises questions regarding the philosophies and methodologies to be used in their synthesis.

Research has been conducted at the University of Wisconsin–Madison since the mid 1960s in the design of computer control systems for manufacturing machinery, cells, and systems. The design of highly distributed manufacturing control systems has been a major area of focus since 1980. This has resulted in the development of design methodologies

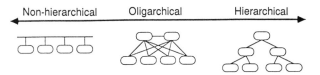

Figure 7.1   *Spectrum of manufacturing cell control architectures*

fault-tolerant, non-hierarchical control of these systems and the construction of a number of experimental manufacturing cells and systems with non-hierarchical controls. The motivation for non-hierarchical manufacturing cell control systems will be described in this chapter, together with the philosophy of their design and examples which include the experimental systems that have been developed.

2   SYSTEM REQUIREMENTS

A desire for greater flexibility, reliability, and performance has led to the application of multiple computer control concepts and network communication concepts to the operation and control of manufacturing machinery, cells, and systems. This development has been enabled by the rapid evolution of the microprocessor which has provided a convenient, cost-effective base upon which advanced manufacturing systems can be constructed. The general practice has been to buy 'attractive-looking' machine and computer components, and then to try to piece them together into a system. The limited success of this approach has made it clear that a more disciplined methodology must be applied in the design of these systems if demands for higher levels of performance, fault tolerance, and cost-effectiveness are to be realized. The late Jozsef Hatvany (Hatvany, 1984) noted that one of the main obstacles to successful design and implementation has been the 'failure to take a top-down, integrated view of the overall system in its early stages' of development. He found that a combination of 'a thorough top-down functional analysis with an ordered bottom-up stream of implementation decisions' can be used 'to design operable, efficient systems on the basis of imperfect and incomplete descriptions' (Hatvany, 1985).

The starting point of the design of a manufacturing cell control therefore is the identification of functional requirements (Duffie, 1982). The nature of the transformation of the cell's inputs to its outputs must be identified along with the speed at which these transformations are to be carried out and the response times required. Only simplistic models may be available at this point, but these models must be of sufficient detail to allow determination of computational and data storage requirements. Suitable representations of components must be developed so that further analysis can be carried out to obtain size and feasibility estimates, operational requirement (reliability, modifiability, extensibility, maintainability, etc.), and criteria for verification of the analysis (Altford, 1979; Palmer, 1979). Interactions with external systems and devices must be specified. The characteristics and requirements of interfaces to human operators, electromechanical systems, and existing computer systems must be known as well as their physical location. The information content of communications with these systems and devices also must be known along with the rates at which they must be serviced. The mechanisms by which failures can occur must be identified along with their probability and means of detection. A specification must also be made with regard to extensibility, and the potential need for future modifications, the nature of which may be unforeseeable at the time of design. These may be lumped under a need for flexibility which may be very important for general-purpose designs, but not as important for single-purpose designs. These considerations lead to the specification of the functional requirements and performance requirements that are to be satisfied.

3   PARTITIONING

Once requirements have been specified, they must be clustered so that they can be assigned to functional entities and computing resources in the control system. The philosophy used in this partitioning process has a significant impact on the final architecture of the control system. As indicated in Figure 7.1, a spectrum of control architectures exists, which varies from rigidly hierarchical to completely non-hierarchical or isoarchical. Systems at various points in this spectrum have been studied and implemented, and will be discussed in the following subsections.

3.1   Hierarchical Control

The hierarchical control architecture is by far the most commonly used architecture, in part because past designers of complex computer-controlled manufacturing systems have been conditioned by the concepts of hierarchic structures whose tree-like organizations are similar to corporate organization structures (Hatvany, 1985). Hierarchically controlled systems are constructed using a philosophy of 'levels' of control and contain a number of control entities arranged in a pyramidal structure (Albus et al., 1986; McCain, 1985; Simpson et al., 1982). Commands input at the highest level are decomposed into more detailed commands and passed on to the next lower level in the hierarchy. Generally, the lowest levels are dedicated to tasks such as machine drive control and position sensing. The top levels control, coordinate, and manage the entire system. Other characteristics of hierarchical systems include: master/slave relationships which exist between levels with command data flowing downwards in the hierarchy

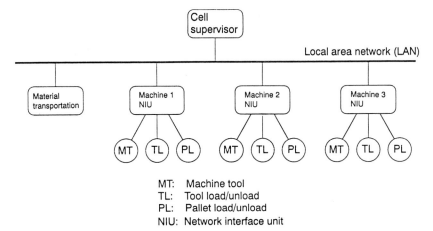

Figure 7.2 *Hierarchically controlled machining cell: MT, machine tool; TL, tool load/unload; PL, pallet load/unload; NIU, network interface unit*

towards machines at the lower levels and sensory data flowing upwards in the hierarchy towards higher management levels; and shared databases which are accessed at higher levels of the structure rather than distributed throughout the system. A system of this type is shown in Figure 7.2 where a high-level machining cell controller interacts with three machine tools via a local area network (LAN). Each machine tool is provided with a 'front-end' or network interface unit (NIU) which integrates its controller with separate tool loading/unloading and pallet loading/unloading subsystems, both of which are serviced by a material transportation system.

The complexity of computer-integrated manufacturing systems with hierarchical architectures tends to grow rapidly with size, resulting in accompanying high costs of development, implementation, operation, maintenance, and modification. The rigid structure of purely hierarchical systems and the tight coupling between master and slave entities usually means that fast response times can be obtained. Experience has shown, however, that the structure of these systems becomes fixed early in their development, making subsequent unforeseen modifications difficult for systems beyond a certain complexity (Duffie *et al.*, 1988; Piche *et al.*, 1983). Extensions must be foreseen in advance, making subsequent unforeseen modifications difficult. An entity at a given level in the hierarchy requires substantial knowledge of the entity above it in the hierarchy as well as the entities below it when fault tolerance is incorporated. This tends to make large hierarchical systems difficult and expensive to design, maintain, and modify. Experience has shown that fault tolerance is obtained in hierarchical systems with considerable expense and complexity.

As an alternative to rigidly hierarchical structure, Piche *et al.* (1983) suggested an oligarchical structure in which higher-level control is vested in a number of coordinators which can communicate with each other as indicated in Figure 7.1. An oligarchical manufacturing system would be a 'multi layered organization whose lowest layer is composed of controllers directing production entities, and any other layer is comprised of co-ordinators'. The controllers would make local decisions for execution of tasks, and where interaction with other controllers is needed, requests for decision and coordination functions would be sent to a coordinator. As illustrated in Figure 7.3, a robotic assembly system proposed by Piche *et al.*, coordinators at a given level can communicate directly or indirectly with all other coordinators at the same level and the next higher level. This allows normal operation in a hierarchical manner, but operation in a non-hierarchical manner when failures in coordinators or communcation links occur.

3.2   Non-hierarchical Control

It is generally accepted that increased autonomy of the participating components of a system reduces the need for a highly intelligent centralized governing body. A system with autonomously functioning components will not collapse when one or more of the components fail or malfunction. Autonomy enforces localization of information, isolating each entity from other entities in the system. Programmers of entities need only be concerned with the logic of the particular entity they are developing and a specified set of well-defined messages to be exchanged between entities. They need not be aware of the inner workings of other entities of the system. It is well known that this approach results in fewer logical programming errors and fewer unforeseen interactions between entities. Unfortunately, non-hierarchical structures do not resemble hierarchical management and production control structures, and traditional deterministic design phil-osophies and methodologies are not necessarily appropriate for non-hierarchical systems. Traditional engineering methods used to develop and implement computer software and hardware for machinery control appli-cations are not necessarily appropriate for the design of these systems. A new design philosophy is required in their synthesis.

Hatvany (1985) stated:

It is time that we freed ourselves of the tradition of equating hierarchical structuring with knowledge and orderliness . . . . Highly centralized and hierarchically ordered systems tend to be rigid, constrained by their very formalism to follow predeter-mined courses of action. However carefully 'optimized' their conduct may be, it has been shown that this very property of inherent resistance to organization change itself necessarily leads in due course to catastrophic collapse.

In the place of a hierarchical structure for manufacturing systems, Hatvany suggested a 'cooperative heterarchy', a non-hierarchical approach in which 'while there are no "higher level" controllers in the system, nevertheless

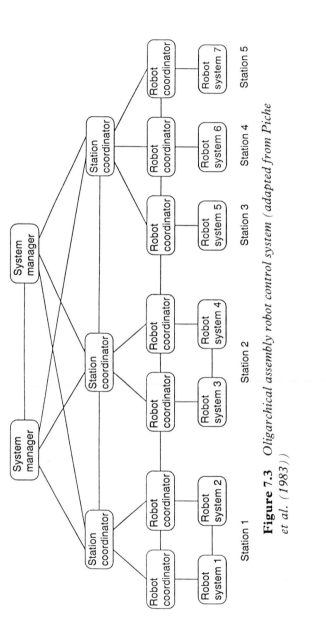

Figure 7.3 *Oligarchical assembly robot control system (adapted from Piche et al. (1983))*

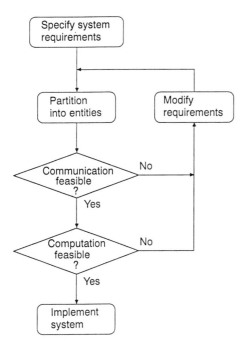

Figure 7.4 *Design of a system of cooperating, communicating entities*

each member must conform to certain rules in order to obtain certain privileges'. The result of a thorough top-down analysis of functional requirements based on imperfect and incomplete descriptions followed by an ordered bottom-up stream of implementation decisions would be 'an internally well-ordered, but externally permissive system' in which:

- entities have equal rights of access to resources;
- entities have equal mutual access and accessibility to each other;
- entities have independent modes of operation; and
- entities strictly conform to the protocol rules of the overall system.

Designers would establish and codify rules, the observance of which by entities in the system is mandatory in order to obtain privileges. Each entity would have a dual set of goals: the first set emphasizing local autonomy, flexibility, and a high degree of local intelligence to cope with unexpected events; and the second set emphasizing optimal overall system operation and survival when subjected to unexpected events.

Duffie advocated the approach illustrated in Figure 7.4 in which system functional requirements are decomposed and partitioned into a set of quasi-independent, communicating entities (Duffie, 1982). The following principles would be applied in the partitioning process:

- There is a natural partitioning associated with the system.
- The result of partitioning is a set of quasi-independent, communicating entities with relatively weak interactions.
- All communication between entities takes the form of messages transmitted on a network.
- The physical system configuration should be transparent to the entities in the system, and entities should not need to know where other entities reside.
- Time-critical responses should be contained within an entity and should not be dependent on time-critical responses from other entities.
- The resilience and viability of individual entities should be a major criteria.

The result of partitioning according to these principles can be a non-hierarchical system in which all entities operate and communicate at an equal level. The entities designed may contain functions that, in a strictly hierarchical architecture, would have to be placed in different entities. The issue here is that communication between entities, fault tolerance, and other considerations may result in a system architecture that is different from any preconceived functional hierarchy.

4   FAULT TOLERANCE

One of the most difficult issues in the design of complex control systems is achieving fault tolerance. This includes the automatic detection of failures, diagnosis of the cause of failure, and the determination and implementation of the appropriate recovery actions. Typically, a hierarchical system will be designed to perform a set of functions with the assumption that all entities will perform correctly. Then, various fault possibilities are identified along with the combinations in which they are expected to occur and their anticipated effect on the system as designed. Additional control logic is then designed that will explicitly detect these faults and allow the system to recover from them. Unfortunately, the complexity of the additional control logic which is required to achieve fault tolerance can be an order of magnitude greater than that required for the system in the absence of faults. This significantly increases the cost of the system as well as the time required for design, implementation, and debugging.

Design principles of eliminating master/slave relationships and minimizing global information enhance the attainment of implicit fault tolerance and significantly reduce the need for additional, explicitly programmed fault tolerance. Hatvany (1985) recognized that 'no algorithms can be written which foresee every possible failure mode of a highly complex system, nor can remedial strategies be deterministically designed for every situation'. If system designers have an incomplete knowledge of the interactions between system entities, it is important to avoid assumptions about the current state or status of other entities in the system when designing an entity. Elimination of this and other forms of global information in a system tends to enhance the following:

- containment of faults within entities;
- recovery from faults in other entities;
- system modularity, modifiability, and extendability;
- complexity reduction; and
- development cost reduction.

Eliminating master/slave relationships and minimizing global information enhances the attainment of implicit fault tolerance and significantly reduces the need for explicitly programmed fault tolerance. Duffie *et al.* (1988) developed a number of design principles that tend to produce systems of cooperating autonomous entities with a high level of intrinsic modifiability, extensibility, and fault tolerance. These principles are as follows:

- Entities should possess the highest achievable level of local autonomy.
- Master/slave relationships should not exist between entities.
- Entities should cooperate with other entities whenever possible.
- Entities should not assume that other entities will cooperate with them.
- Entities should delay establishing relationships with other entities for as long as possible.
- Entities should terminate relationships with other entities as soon as possible.
- Information generated by an entity should be retained locally rather than globally.
- Entities receiving information from other entities should not assume that the information is correct.

Many of these principles can be traced back to the basic principle of minimizing global information. Global information can be considered to be any information that is not local to (confined to) a single entity. This includes any assumptions made by an entity about the current state or status of another entity. It is the very presence of assumptions about the status of entities in a complex system and their ability to perform required functions that makes fault tolerance so difficult to achieve. System designers often have incomplete knowledge of system components and component interactions, and system requirements may be so complex that the system becomes undesignable or unimplementable with traditional techniques. Furthermore, implemented systems may be unmodifiable because of high levels of complexity. The presence of global information and complex relationships between entities makes modification and extension expensive, prone to the introduction of logical errors, and often not achievable in the field.

At the most elementary level, faults should be detected and diagnosed within entities, allowing them to undertake individual recovery actions. If these individual actions are reasonable globally, all entities in the system will converge to an acceptable new state after all interactions have completed their course. However, some classes of faults may require distributed problem solving conducted jointly by one or more entities in the system. Actions taken locally within entities must be globally coherent, and this global coherence must be achieved by local actions alone when there is minimal global information and an absence of master/slave relationships. According to Cammarata *et al.* (1983), the key to achieving fault tolerance in a distributed system lies in the fact that 'while distributed agents have greater difficulties in solving a given task, they have potentially more options as well'. For example, an individual entity may act in response to detecting a fault, but may also request other entities to do so.

A number of experimental, non-hierarchical manufacturing cell and machine control systems have been developed at the University of Wisconsin–Madison. Global information was minimized in these systems as was the complexity of relationships between entities. The remainder of this chapter illustrates, in several examples, the design philosophy which has been developed. The discussions of these examples contain further information regarding the trade-offs of hierarchical and non-hierarchical strategies, as well as results obtained from comparisons of system implementations. They show that the non-hierarchical philosophy and architectures that have been developed can be used successfully to design and implement manufacturing control systems in which fault tolerance, modifiability, extensibility, and reduced complexity are of high importance.

5   NON-HIERARCHICAL CONTROL OF MACHINE DRIVES

A non-hierarchical control was designed and implemented at the University of Wisconsin–Madison which allowed configuration and control of a large number of machine drives, with various position relationships maintained between them (Duffie, 1982; Duffie and Bollinger, 1983). As illustrated in Figure 7.5(a), drives could be configured as a master drive, with the position of other drives slaved to its position. Figure 7.5(b) illustrates another possibility in which drives could be configured to operate independently. Because the configuration could vary from application to application, a generalized system was desired in which any number of drives and a variety of relationships between drives could be specified to suit the requirements of a given application. Multiple-computer control was indicated because of the large number of drives to be controlled, and the

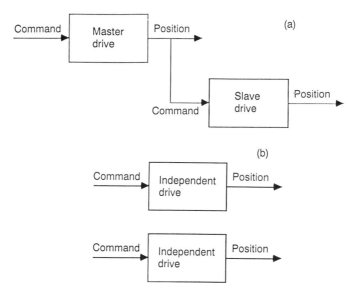

Figure 7.5 *Master/slave (a) and independent (b) drive configuration*

potential for significant physical distances between drives. Examples of applications for this type of control system included: synchronization of stations in a flexible assembly cell where stations must be indexed and controlled with respect to the position of other stations in the cell; sheet material handling systems where tension must be controlled between successive sets of pinch rollers; and robotic manipulators where end-effector motion depends on coordinated motion of multiple axes.

Figure 7.6 shows the functional organization of the general-purpose, multiple-drive control system which was developed. The basic functional requirements for the system included the following:

- The number of drives to be controlled was to be transparent to the system.
- The number of control computers and the mix of hardware (various makes and models) was to be transparent to the system.
- The configuration for a given application was to be determined by the user by selecting the number of drives to be controlled and the interrelationships between drives.
- Position commands were to be generated for master drives from prestored motion profiles.
- Slave drives' position commands were to be obtained from the position of a specified master drive.
- All drives were to be closed-loop controlled using position feedback.
- Faults in any given drive were to be isolated as thoroughly as possible from other drives.

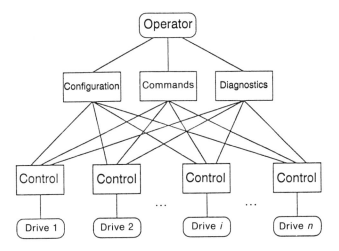

Figure 7.6   *Configurable drive functional diagram*

Following a 'top-down' design approach, the requirements of the system were completely defined before the system was partitioned into entities and implemented. When the requirements of the system had been defined, the multiple-drive control system was partitioned into processes. Of the functional requirements given above, some, such as closed-loop control of individual drives, naturally fall into independent entities whereas others, such as master/slave drive configuration, tend to represent relationships between entities rather than an entity itself. With the design principles and the functional requirements of the multiple-drive system in mind, the tasks to be performed by the system were partitioned as described in the following subsections.

5.1   Drive Control Entity

It is advisable to start with the most obvious portions of the partitioning first and then proceed with the remainder of the system. The principle of containing time-critical responses within an entity implied that closed-loop control for each drive should be implemented as one entity. This conclusion was re-enforced by the fact that the sample rates normally required to control a drive would require excessively high communication rates through the network in order to minimize time delays if control loops were divided between entities. A drive control entity was therefore defined. Commands for the drive could be either generated by the entity if configured in the master mode, or obtained by communicating with a specified master drive in the system if the drive was configured to operate in a slave mode.

More than one drive could have been controlled by an entity. This would have reduced system communication requirements, particularly if there was *a priori* knowledge of master/slave relationships between drives; however, this was not possible in a general-purpose system where no *a priori* knowledge of configuration was assumed. A separate drive control entity was therefore used for each drive. Following the principle that information was to be retained locally rather than globally, all configuration information for a given drive was contained wholly within its drive control entity. No information was stored in master drives regarding which drives were slaved to them. Slave drives communicated directly with master drives, requesting and receiving position information. Fault isolation and tolerance was achieved because failure of a given drive was confined to only that drive. Slave drives detected failure of their master drive when the master drive failed to respond properly to requests for position information. The slave drives then undertook recovery actions to allow them to survive the fault in a degraded state.

5.2   Operator Entity

Keyboard and display input/output with the user formed the basis of an operator entity. Following the principle of hiding information from other entities, an operator entity was defined that was capable of handling all communication with the user during the configuration and operation of the system. The operator entity then communicated when necessary with the drive control entities in the system. Configuration information regarding relationships between drives was stored in slave drive control entities rather than the operator entity, eliminating global information. The resulting system was very flexible in terms of configuration, modification, and extension. Fault tolerance was increased because any number of operator entities could exist, and even if all the operator entities failed after the system had been configured, the drive control entities could continue to perform their master and slave functions. This feature required no additional software development because of the high level of autonomy of the entities in the system and the minimization of global information.

5.3   Implementation

The result of partitioning was a configurable, fault-tolerant system design with the non-hierarchical architecture shown in Figure 7.7(b) rather than the traditional architecture shown in Figure 7.7(a). A communication network was selected which had sufficient communication rates and reliability to transmit position information between drive control entities. The system was capable of supporting an arbitrary number of drive control entities. One drive control entity was allocated to each control computer, allowing all control software to be written in a high-level language rather than assembly language. Each of the control computers could also support an operator entity, allowing redundant operator interfaces to exist if desired.

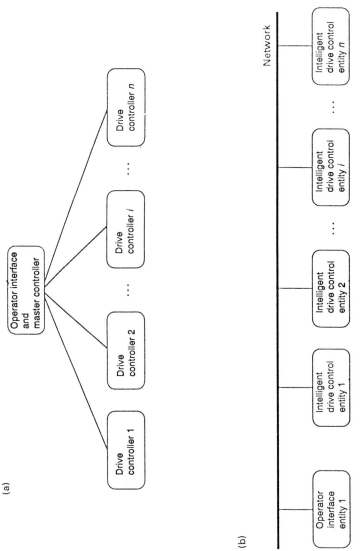

Figure 7.7 *Hierarchical (a) and non-hierarchical (b) architecture for multiple-drive control system*

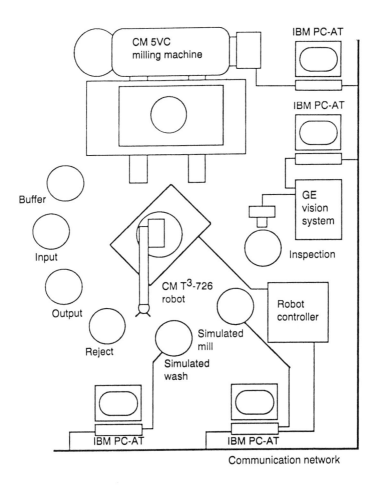

Figure 7.8 *Experimental flexible machining cell*

6   CONTROL OF A MACHINING CELL

Three machining cell control systems were implemented at the University of Wisconsin–Madison in an effort to evaluate the feasibility, advantages, and disadvantages of non-hierarchical cell control (Duffie *et al.*, 1986; Duffie and Piper, 1987). A cell consisting of a machining centre, vision inspection system, and material handling robot was constructed first. This cell was controlled using a single central computer. The new cell illustrated in Figure 7.8 was then constructed. This cell consisted of the machining centre, material handling robot, and inspection station from the previous cell, together with an input station, output station, reject station, buffer station, simulated wash station, and a simulated second machining centre. Initially, a hierarchical cell control was implemented using four micro-computers. This hierarchical cell control was then replaced with a non-hierarchical cell control. Details of the centralized, hierarchical, and non-hierarchical cell control systems eveloped are given in the following subsections.

6.1   Centralized Control

A centralized cell control using a single computer and a lock–step scheduling algorithm for part/machine sequencing was the first to be constructed. Only two part types could be processed, and only one part could be in the cell at a time. To initiate the processing sequence, a part was placed at a the cell's input/output station. The system then cycled through an explicitly programmed set of states while processing each part. The processing steps for each part were hard coded into control software. This system could not handle machine and communication failures, and could not be modified to accommodate additional stations or parts without substantial rewriting of control software.

6.2   Hierarchical Control

A hierarchical cell control using a supervisory computer, three slave computers, and a dynamic scheduling algorithm for part/machine sequen-cing was the next to be constructed. A list of all stations in the cell was maintained by the supervisory computer, and a database was supplied to the supervisor with each part specifying the steps required to manufacture the part. Scheduling was performed in real time to move parts to the next station specified in their database when three conditions were satisfied: processing was completed at the current station, the next station was ready to accept a part, and the robot was available to transport the part to the next station. This hierarchical cell control strategy exhibited improved flexibility and fault tolerance when compared with the centralized system. This was due to the distribution of low-level control functions to separate control com-puters. Unfortunately, supervisor software and databases had to be modified when stations were added or deleted from the cell.

6.3   Non-hierarchical Control

The final control alternative considered was a part-oriented non-hierarchical control where each station and part was programmed as an individual 'intelligent' software entity, running under a multi-tasking operating system. This development was a direct result of the desire to eliminate global information and to locate the decision making where the information originated, hence eliminating the supervisory computer. Every

Table 1 *Controller cost and performance comparison*

	Centralized controller	Hierarchical controller	Non-hierarchical controller
Lines of source code	680	2450	1235
Software development cost[a]	$17 000	$61 250	$30 875
Expansion software cost[b]	$17 000[d]	$960	0
Machine utilization[c]	—	64%	60%
Complexity	low	highest	lowest
Flexibility	lowest	high	highest
Modifiability	lowest	high	highest
Fault tolerance	lowest	low	highest
Intelligent parts	no	no	yes

[a] At $25 per line of source code.

[b] 24 hours per machine added at $40 per hour.

[c] From simulated fault-free cell operation, random part mix.

[d] Complete redevelopment required due to explicit sequencing.

functional entity in this control system used the communication network to arrange transactions with other entities. For example, part entities in the system negotiated with station entities in the system to make reservations for input, output, machining, and inspection. Part processing sequence information was confined to the part entities, and the status of processing stations was confined to the station entities. No entity in the system had complete knowledge of the state of the system or a production schedule. As a result, entities continued to function when failures in other entities occurred, and stations could be added to the cell without modifying the software of other entities.

6.4   Cost and Performance

Cost and performance data collected in these cell control developments is shown in Table 7.1. The software for the non-hierarchical cell control contained fewer lines of source code than the hierarchical system. This indicated that the non-hierarchical cell control was significantly less complex than the hierarchical system (the number of source code lines was used as an indicator of software complexity (Gremillion, 1984)). It also indicated that software development costs were significantly less for the non-hierarchical cell control. While it had been expected that the complexity of the station entities would be greater with no cell supervisor, it was observed that the complexity of the logic required to allow station entities to communicate with part entities was no greater than that required between the stations and the supervisor in the hierarchical cell control. The supervisor and all of its complexity were therefore eliminated without a significant accompany increase in station software complexity. Part entity software was very simple, and replaced the part processing databases used in the supervisor of the hierarchical system.

It should be noted that an operating system supporting network communications and multi-tasking operating systems was required for the non-hierarchical cell control. Although it was more complex than the operating system used for hierarchical control, it was generic and contained no manufacturing-oriented databases and no cell control features. Cell control application software was written as sets of autonomous Pascal programs that established cooperating part and station 'intelligences' which communicated via straightforward message sending and receiving procedures. The complexity of the operating system was hidden from the cell control software developers who were only required to produce simplified, localized entity software. This resulted in low development costs for cell control software, and also resulted in low modification and expansion costs during later evolutionary changes in the system. An additional benefit of this approach was the ease with which decision-making logic could be incorporated into part processing sequences. This capability was used for reacting to results obtained from inspection (accept, reject, rework, etc.) (Duffie *et al.*, 1986), and also could have been used for selecting alternative processing sequences based on tooling availability (Ranky, 1986).

Run-time memory requirements were lower for the non-hierarchical system, and its distributed nature tended to balance CPU loads and utilization. Long-term machine utilization in the cell, under normal operation without faults, was predicted from simulation results to be lower for the non-hierarchical system because of the necessity for cooperative scheduling using network communications. However, because faults occur regularly in manufacturing systems (Yak *et al.*, 1983), it is expected that the improved fault tolerance of non-hierarchical systems will counteract this trend.

7   MULTIPLE-CELL CONTROL

An experimental, non-hierarchically controlled, flexible manufacturing system for automated production of moulds for plastic processing was constructed at the University of Wisconsin–Madison. The system consisted of the hardware illustrated in Figure 7.9 and contained a machining cell, an EDM cell, a CMM cell, and a CAD/CAM cell. The system was required to support the following functions:

- CAD design of plastic part and mould geometry;
- machining of mould cavities;
- finishing of mould cavities;
- inspection of mould cavity geometry;
- inspection of plastic part geometry; and
- modification of mould cavity geometry.

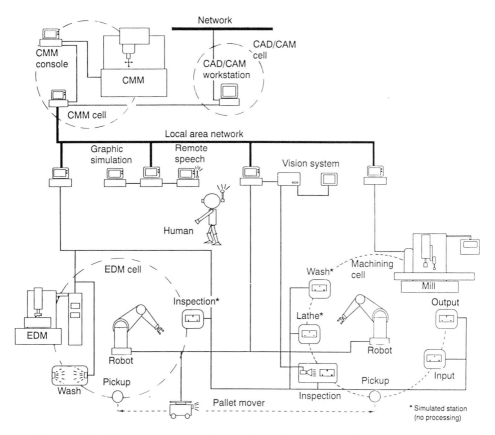

Figure 7.9 *Experimental mould manufacturing system with non-hierarchical control*

The design principles described earlier were applied in the partitioning of this system into a set of functionally heterogeneous, communicating entities. These included:

- mould entities;
- machine entities (mill, wash, inspection, EDM, CMM);
- material handling entities (robot, pallet, mover);
- CAD/CAM entities;
- geometric database entities;
- human entity; and
- input and output station entities.

Parts to be manufactured were programmed as 'intelligent' entities that cooperatively interacted with 'intelligent' robots and processing machines to satisfy system production requirements. Human entities were also included, and voice communication capabilities were implemented to allow them to cooperate conveniently as colleagues of other entities in the system.

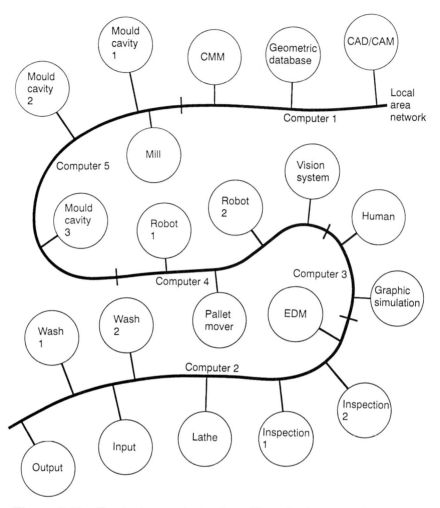

Figure 7.10 *Entries in experimental mould production control system*

These entities were distributed over five microcomputers on a local area network as shown in Figure 7.9. The production resources which the entities controlled in the system were connected to the microcomputers as indicated in Figure 7.10. It should be noted that these connections were grouped for convenience in implementation rather than by cell. The traditional cell and cell supervisor concept did not exist in this control system. In fact, the logical connections between entities are better represented by the single network of entities shown in Figure 7.10 than by the cells shown in Figure 7.9. In this system, the geographic location of physical resources and physical relationships between these resources was transparent to the microcomputers and entities they supported.

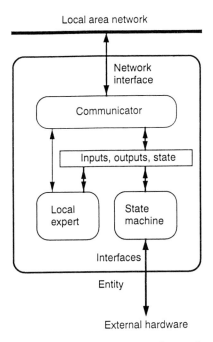

Figure 7.11   *Structure of generic system entity*

Because of the heterogeneous nature of the entities in the system, the generic entity shown in Figure 7.11 was developed which allowed the entities and their interconnections to be handled in a homogeneous fashion. Each entity consisted of three components: a communicator which handled communication with other entities in the system, a state machine which implemented the control logic of the entity and interacted with external hardware associated with the entity, and a local expert. In the following subsections, the concept of a generic entity is introduced, and the various types of entities found in the experimental system are defined. Transactions between entities are then discussed together with the reservation strategy used for dynamically establishing and terminating relationships between entities.

7.1   Communicator

All of an entity's network communication functions were performed by the communicator. The communicator allowed the entity to asynchronously exchange messages with other entities via the communication network. The communicator received messages related to transactions with other entities, fault announcements, and requests for information on the state of the entity. Depending upon the message received, functions were performed such as changing the communicator inputs to the state

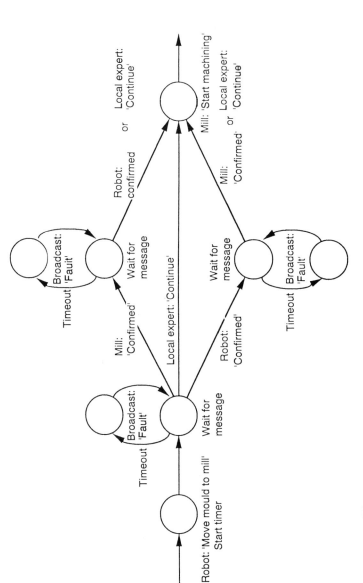

Figure 7.12 *Portion of the state transition diagram for a mould entity*

machine or sending the state of the entity to another entity. The communicator watched for specific state machine and local expert outputs which required generation and transmission of messages relating to transactions between entities, fault announcements, etc. It then transmitted the required message. The communicator was event driven in the sense that the transmitting or receiving of a message from another entity was treated as a spontaneous event which was immediately responded to and then forgotten.

7.2   State Machine

Each entity passed through a programmed sequence of states as it performed its required functions. All allowable state transitions were defined as functions of communicator inputs, external inputs, timers, etc. In this regard, it resembled a programmable logic controller. These transitions were represented by a state transition diagram which served as a reference in both the design of other entities and in the eventual operation and maintenance of the system. The state transition diagram in Figure 7.12 illustrates an example in which a mould entity required verification of placement of the mould on a machine tool by a robot. Placement was verified by messages from both the mill entity (which had limit switches incorporated into mould fixturing hardware) and the robot entity (which detected release of the mould by its gripper). These messages were received by the communicator and used to set appropriate inputs to the state machine. Failure to receive these messages within a required length of time resulted in a fault announcement being broadcast by the communicator when the state machine set an appropriate output. Additional state transitions were included which allowed the part entity to proceed if allowed to by its local expert.

The communication network was used by each entity in the system to negotiate transactions dynamically with other entities for purposes of real-time scheduling of processing, transportation, etc. To illustrate this concept, consider the case of a mould that was to progress from the input station to a milling station to a washing station and finally to an output station. Each of these movements was performed by a material handling robot. When the mould arrived at the input station, the first step of the mould entity was to broadcast a message requesting a milling station. Any milling machine entity that was free to process the mould could respond by sending a message to the mould entity indicating its availability. The mould entity then returned a message to a responding milling machine entity asking for a reservation. If a milling machine entity acknowledged this reservation, then the mould entity proceeded to communicate with the robot entity to arrange its transportation from the input station to the milling station. Otherwise, the mould entity would broadcast again to request a milling station. Figure 7.13 illustrates a case where there were three milling stations in the machining cell (one real, two simulated). Mill 2 was busy, but

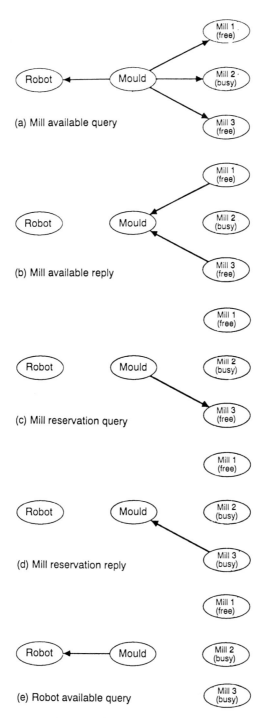

Figure 7.13   *Reservation of a mill entity by a mould entity*

Mill 1 and Mill 3 responded to the request broadcast by Mould. If the message from Mill 3 was received first, Mould sent a reservation message to Mill 3. If Mill 3 had not, in the meantime, made a reservation with another mould entity, it then acknowledged Mould's reservation. This established a relationship so that the mould could be transported by the robot to Mill 3 and subsequently be processed.

Milling machine entities were not required to respond to any messages from mould entities, nor were mould entities required to act upon responses from milling machine entities. For example, if responses were received from more than one milling machine entity, all but one was ignored. Similarly, a milling machine entity could respond to requests by many moulds for its services, but could ignore reservation messages that were received after a reservation has been made. This reservation transaction process resulted in a self-configuring and self-reconfiguring system. In the case in Figure 7.13, Mill 3 was implicitly included in the system if it was able to communicate with and make reservations with other entities. If Mill 3 failed or was physically removed from the system it no longer responded to messages such as those generated by Mould, thereby implicitly excluding itself from the system. No additional software or databases were required anywhere in the system to obtain this modifiability and fault tolerance.

7.3   Local Expert

The local expert was created to support advanced capabilities such as distributed problem solving and real-time scheduling (Cammarata et al., 1983; Fox and Kempf, 1985; Shaw and Whinston, 1985). It was included to allow exchanges of messages with other entities through the communicator for purposes of fault detection, diagnosis, recovery, and reporting. It also provided inputs to the state machine for triggering appropriate fault-recovery functions. The local expert could perform local problem solving as well. In the example shown in Figure 7.12, the Mould entity's local expert could allow its state machine to continue after detecting and diagnosing a fault in which the Mill and/or the Robot incorrectly failed to confirm the movement of the mould to the Mill. Machining could be allowed to begin after the Mould entity's local expert confirmed the presence of the mould on the milling machine, perhaps after communication with the Human entity.

7.4   Human Entity

As indicated in Figure 7.9, a Human entity was integrated into the system using remote speech capabilities which included voice recognition, speech synthesis, and remote communications. A graphic simulation provided the human with visual information regarding the state of the system. In the Human entity, the human being functioned as the local expert. Messages announcing faults were occasionally broadcast by other entities, received by the Human entity, and passed to the human via remote communications and speech generation. The human could, via speech recognition, transmit advice to the local expert of another entity, with the entity receiving the advice having the option of responding to it if it was programmed to do so. The Human entity was not required to respond to messages from other entities, and these entities did not assume that it would respond to them. Hence, an ambulatory human was integrated into the system as a colleague capable of conversing with other entities in the system, observing their state, and giving advice which the entities could act upon if desired. The human could therefore act as a provider of complex fault diagnosis and recovery functions, while maintaining the local autonomy and hence the resilience, viability, and fault tolerance of other entities.

7.5   Role of Simulation

Although the cells in the manufacturing system contained a number of simulated stations, the other entities in the system could not distinguish between whether they were communicating with a real or a simulated station. The only difference was that manufacturing processes could not be performed at the simulated station, and the physical moulds on the pallets left the station in the same state in which they arrived. In fact, it was possible to replace a real station by a simulated station at any point in the operation of the experimental system, and to create additional simulated stations at will so that more complex systems could be studied. This capability was a natural consequence of local autonomy and information localization in the system. It allowed the system to be constructed entity by entity, and initially assembled and debugged using completely simulated entities. The material handling robot entities were then modified to control the actual robots, followed by the machine tool and other entities. In this way, the system software was evolved in a step-by-step manner from simulated station control to real station control. This system development process can be summarized as follows.

(1) Construct initial system using simulated stations.
(2) Operate and debug system using simulated stations.
(3) Add machine interfaces to a station.
(4) Operate system with newly interfaced station.
(5) Repeat (3) and (4) until all stations have been interfaced.
(6) Operate and debug system with actual rather than simulated stations.

Proposed additions to a system could be implemented as simulated entities, and the results studied prior to bringing in the actual hardware. The ability to simulate entities was used to expand the experimental system to include more processing entities than actually existed in processing hardware. It should be noted that the same microcomputers and operating system were used for both simulated station control and real station control, and a given entity could be switched back and forth whenever desired between simulation and reality. The timescale in the system could be modified either to make time appear to pass more slowly to aid in observing interactions between entities, or to pass more quickly to aid in studying the operation of the system as a whole. The latter was most useful in the early stages of development when all entities in a system were simulated, but also could be used at any time to study alternative system configurations and long-term system operation.

8   CONCLUSION

Non-hierarchical control is an attractive alternative for future manufacturing cell control systems. The presence of multiple independent, communicating, problem-solving 'intelligences' in a system creates significant new opportunities in manufacturing system design that go beyond traditional deterministic approaches (Hatvany, 1978, 1983). It is clear that non-hierarchical control systems consisting of fifty to eighty entities can be designed and built. The need exists for further advances in design and implementation of larger and more complex manufacturing systems so that the objectives of higher performance, lower development costs, improved modifiability, and higher fault tolerance are achieved. The existence of 'intelligent' parts or workpieces may change our view of the roles of software and databases in manufactuing systems. For example, a part designer can incorporate sophisticated inspection, correction and rejection software into a part entity. Parts then can be 'self-inspecting' and 'self-documenting' using locally stored information collected during program execution. Finished parts can be programmed as 'users' of other parts, or programmed to 'serve' other parts in the assembly of a final product. It may be unnecessary to support complex databases for production control when part entity software can be either generalized to produce batches of parts or individualized to produce one of a kind.

A number of important aspects of non-hierarchical system design methodology presented here and the experimental systems in particular should be emphasized. First, fault tolerance and modifiability are implicit in the design. The principle of establishing all relationships between entities using network communications ensures that entities are insulated from each other and can be constructed to function in a locally autonomous manner. A strategy of 'falling back' to well-defined states in which progressively fewer relationships exist between entities ensures that the system is well behaved even when major faults occur, regardless of whether faults are caused by hardware failures or design flaws in software. New entities can be implicitly included in a non-hierarchical system, with no need to reprogram any other entities in the system, by virtue of their ability to communicate. Also, additional processors supporting additional entities can be attached at any time to the communication network, again without a need to reprogram other system entities. Similarly, entities are implicitly excluded from the system when they cease to communicate. This can be due to a fault in the entity or its physical removal from the system. This implicit modifiability and reconfigurability can be an important benefit of this non-hierarchical design.

In hierarchical structures, the jobs performed by human beings are mostly either manual tasks or supervisory and decision-making functions in abnormal situations when control of part or the whole of a system is taken over by human staff (Hatvany, 1985). Unfortunately, the complexity of systems is reaching the level where human operating staff are not able to supervise whole systems, nor are they able to run them in a manual mode. The successful integration of human entities into non-hierarchical systems can take advantage of human deductive reasoning and problem-solving capabilities by allowing them to cooperate in distributed problem solving with local experts in other system entities. The human should not be required to respond quickly, but instead to consider the status of the system, obtain visual knowledge of its physical state, and give advice as appropriate. An ambulatory human being can be integrated into a system as a colleague, observing the physical status of its various entities, conversing with the entities to determine their logical status, and giving advice which the entities can consider and act upon if desired, acting as a primary provider of fault diagnosis and recovery functions.

A number of significant questions remain which can only be answered through future research and implementation experience. The performance of commercially available communication networks (particularly when the network software overhead is considered) may not be adequate to satisfy capacity and response time requirements of large-scale systems. Scheduling has been distributed in our experimental systems, but not optimized in real time. Local decisions made by entities are not globally optimal, and real-time optimization entities may need to be developed that can collect information in a non-hierarchical system and influence the operation of other entities without taking control. Deadlock avoidance is also an important issue that should be addressed. Current scheduling practices in industry result in preplanned production schedules which cannot compensate for failures and other 'surprises' in real time. Appropriate heuristics and distributed, real-time optimization algorithms need to be developed for non-hierarchical systems which do not rely on global information, do not create master/slave relationships, and do not overload communication networks.

The ability to integrate simulation into a system's development is an attractive attribute. It is important to re-emphasize that the same operating system and computers were used for simulation of machinery control as well as for control of actual machinery in the experimental system. Simulations can be integrated with real system control, and proposed additions to a system can be implemented as simulated entities and the result studied prior to bringing in the actual hardware.

Complexity, cost, and performance measurement and evaluation procedures must be developed so that quantitative comparisons can be made between various architectures and philosophies. New methods are required which can be used to compare quantitatively the advantages and disadvantages of non-hierarchical structures with respect to more traditional hierarchical structures. Comparisons can be made of software development cost, but quantitative measures of complexity, functionality, flexibility, and fault tolerance that can be conveniently evaluated and are not 'coloured' by variations between individual designers have yet to be developed. A goal would be to provide the system designer with a means to select the location in the spectrum of architectures in Figure 7.1 that results in an optimal balance between cost, fault tolerance, and performance. Full-scale tests will need to be conducted in industrial settings and with large-system simulations to quantify further the magnitude of anticipated benefits.

It is safe to say that the existence of multiple, communicating, independent 'intelligences' in a system offers new opportunities for system control and organization. It has been shown that non-hierarchical systems for automated manufacturing can be designed and implemented, and that human beings can be successfully integrated into these systems. The design principles that have been presented offer a means by which the important benefits of low cost, low complexity, and high fault tolerance can be obtained. As future automated systems increase in scope and complexity, the non-hierarchical approach has attractive properties that should be given consideration in future system designs in the hope that the controls for these systems can be made designable, affordable, modifiable, extendable, and maintainable.

REFERENCES

Albus J. S., McCain H. G. and Lumia R. (1986). *NASA/NBS Standard Reference Model for Telerobot Control System Architecture (NASREM)*. Technical Report, National Bureau of Standards, Robot Systems Division, December 1986.

Altford M. (1979). Requirements for distributed data processing design. *IEEE Proc. 1st Int. Conf. on Distributed Computer Systems*, October 1979, pp. 1–14.

Cammarata S., McArthur D. and Steeb R. (1983). Strategies of cooperation in distributed problem solving. *Proc. 8th Int. Joint. Conf. of AI*, Karlsruhe, West Germany, vol. 2, 767–70.

Clark D. D. and Svobodova L. (1980). Design of distributed systems supporting local autonomy. *IEEE COMPCON*, San Francisco, CA, 25–28 February 1980, pp. 438–44.

Duffie N. A. (1982). An approach to the design of distributed machinery control systems. *IEEE Trans. Ind. Appl.*, **IA-18**, 435–42.

Duffie N. and Bollinger J. (1983). Development of a menu-driven programmable system of machinery drives. *Ann. CIRP*, **32**, 281–5.

Duffie N. A. and Piper R. S. (1987). Non-hierarchical control of a flexible manufacturing cell. *Rob. CIM*, **3**, 175–9.

Duffie N., Chitturi R. and Mou J. (1988). Fault-tolerant heterarchical control manufacturing system entities. *J. Manuf. Syst.*, 7, 315–28.

Duffie N. A., Piper R. S., Humphrey B. J. and Hartwick J. Jr (1986). Hierarchical and non-hierarchical manufacturing cell control with dynamic part-oriented scheduling. *Proc. NAMRC-XIV*, Minneapolis, MN, 28–30 May 1986, Society of Manufacturing Engineers, Dearbon, MI, pp. 504–7

Enslow P. H. (1978). What is a distributed computing system? *IEEE Comput.*, January, pp. 13–21.

Fox B. and Kempf K. (1985). Complexity, uncertainty, and opportunistic scheduling. *2nd IEEE Conf. of AI Applications*, Miami Beach, FL, pp. 487–92.

Gremillion L. L. (1984). Determinants of program repair maintenance requirements. *Commun. ACM*, **26**, 826–32.

Hatvany J. (1983). Dreams, nightmares and reality. *Comput. Ind.*, **4**, 109–14.

Hatvany J. (1984). Some approaches to integration. *Rob. CIM*, **1**, 227–30.

Hatvany J. (1985). Intelligence and cooperation in heterarchic manufacturing systems. *Manuf. Syst. (Proc. CIRPS Semin.)*, **14**, no 1, 5–10.

Hatvany J. and Nemes L. (1978). Intelligent manufacturing systems, a tentative forecast. In *Preprints of the 7th World Congr. of IFAC, A Link Between Science and Application of Automatic Control, Helsinki, 1978* (Niemi A., Wahlstrom B. and Virkunnen J., eds). Oxford: Pergamon Press, pp. 895–9.

Kleinrock L. (1985). Distributed systems. *Commun. ACM*, **28**, 1200–13.

McCain H. G. (1985). A hierarchically controlled, sensory interactive robot in the Automated Manufacturing Research Facility. *IEEE Int. Conf. on Robotics and Automation*, St Louis, MO, 1985, pp. 931–9.

Palmer D. F. (1979). Distributed computing system design at the subsystem/network level. *IEEE Proc. 1st Int. Conf. on Distributed Computer Systems*, October 1979, pp. 22–30.

Piche P., Kreiger M. and Santoro N. (1983). Oligarchy, a control scheme for flexible manufacturing systems. *2nd IASTED Int. Symp. of Robotics and Automation*, Lugano, Switzerland, 1983, pp. 35–9.

Ranky P. G. (1986). Tool management tasks in flexible manufacturing systems. *Proc. NAMRC-XIV*, Minneapolis, MN, 28–30 May 1986, pp. 508–15.

Shaw M. and Whinston A. (1985). Task bidding and distributed planning in flexible manufacturing. *2nd IEEE Conf. of AI Applications*, Miami Beach, FL, 1985, pp. 184–9.

Simpson J. A., Hocken R. J. and Albus J.S. (1982). The Automated Manufacturing Research Facility of the National Bureau of Standards. *J. Manuf. Syst.*, **1**, 17–23.

Vamos T. (1983). Cooperative systems – an evolutionary perspective. *Control Syst. Mag.*, August, pp. 9–14.

Yak Y. W., Dillon T. S. and Forward K. E. (1983). Incorporation of recovery and repair time in the reliability modelling of fault-tolerant systems. *Proc. 3rd IFAC/IFIP Workshop on Safety on Computer Control Systems*, Cambridge, UK, September 1983, pp. 45–52.

8   Representation of the cell control task

Paul Rogers

1   INTRODUCTION

This chapter addresses the definition of the decision-making logic which is needed to control manufacturing cells. It begins with a brief discussion of the challenge facing manufacturing system designers and identifies the goals of responsiveness and integration. Structural options for manufacturing systems design are surveyed and the need, independent of the system architecture, for more intelligent decision making is established to allow these goals to be attained. This emphasizes the important role played by 'cell control' and leads to a focus on appropriate modelling techniques for the making of real-time, state-driven control decisions. With cell control centre stage, the requirements of representations for the modelling of manufacturing cells and cell control decisions are examined. A number of alternative techniques which have received attention from researchers are briefly introduced. One particularly promising approach, that of object-oriented modelling, is described in greater detail. The chapter closes by emphasizing some shortcomings of the conventional object-oriented programming paradigm that must be tackled by further research.

2   THE CHALLENGE FOR INTELLIGENT MANUFACTURING SYSTEMS

In a globally competitive market for products, manufacturers are faced with an increasing need to improve their responsiveness to the demands of their customers in order to survive. Such responsiveness involves two requirements: the development of systems which can respond more effectively to changes in their environment and the creation of a capability to design and implement such systems quickly.

The achievement of both of these objectives is critically dependent on the development of appropriate modelling techniques to enable the rapid design of intelligent manufacturing systems that are better equipped to make decisions on the basis of their current state.

Manufacturing systems are complex, dynamic, and stochastic entities consisting of a number of semi-independent subsystems interacting and intercommunicating in an attempt to make the overall system function profitably. The characteristics of this type of system which make the problem of system control particularly complicated stem from: the quantity of data to be handled within the system; the uncertainty of the environment of the system where many disturbances may occur; the structure of the system and the complex relationships between the interacting semi-autonomous subsystems which are part of it.

Manufacturing systems have often been viewed as hierarchies consisting of a number of levels each of which operates according to its own timescale. Improved responsiveness with this view implies that the decision-making capacity of each level must be enhanced such that choices are made on the basis of a wider, more appropriate range of information describing the current state of the system. A system that has sufficient responsiveness to adapt its behaviour automatically to a wide range of circumstances can be considered *intelligent*. Intelligent manufacturing does not mean solely that each subsystem has considerable decision-making power and can respond quickly to the disturbances facing it, but additionally that the subsystems must be so designed that their coordinated behaviour achieves the overall objectives of the system.

One of the overriding factors to bear in mind is the dynamism of the manufacturing environment which ensures that the only certainty is that it will be necessary to change or adapt the behaviour of the system in some way in both the medium and the long term (in addition to providing the system with the capability for short-term adaptive response). True flexibility is thus required in the sense that the system should be able to respond to all the probable situations in which it may find itself. Not only must operational systems be responsive but additionally they must be easy to change. This implies that better and faster design and analysis techniques are required to allow the system design to become as responsive as the product design has been forced to become by market pressures (faster design-to-operational life cycles are required for processes as well as products).

3   THE DESIGN OF MANUFACTURING CELLS AND SYSTEMS

As noted above, manufacturing systems consist of a large number of interacting entities all making decisions (or carrying out actions) which affect the performance of the overall system. It is widely accepted that the decentralized decision makers comprising a manufacturing system need to coordinate their activities in some way to achieve the overall goal which Ruff (1985), for example, has termed *manufacturing systems integration*.

Integration does not simply mean establishing communications links between the diverse hardware and software elements comprising the system, but a deeper cooperation between the different manufacturing functions (see Rogers *et al.*, 1990a) through the sharing of information on the products and processes involved. The representation of this critical information is key to the achievement of true integration.

3.1   A Hierarchical Manufacturing Systems Architecture

It has been claimed by Singh (1980) that a hierarchy is the most natural way to structure the control of a large and complex system whilst others, such as Gershwin *et al.* (1986) have gone further, regarding hierarchical decomposition as offering the only hope of controlling such systems effectively.

Most approaches to the design of manufacturing control systems involve the decomposition of the decisions to be made into a number of hierarchical levels, although the approaches differ in the number of distinct levels identified. Gershwin *et al.* (1986) work with three levels, Bourne and Fox (1984) with four, whilst perhaps the best-known body of research literature, describing the architecture of the Advanced Manufacturing Research Facility (AMRF) of the National Institute of Standards and Technology (NIST) in the USA (see McLean (1985) for a good description), considers five distinct levels. The five levels as defined by the workers at the NIST (e.g. Albus *et al.* (1984)) show the form of this approach.

- *Facility control*, involving manufacturing engineering functions (design of parts and tooling, process planning), information management functions (accounting, sales and purchasing), and production management functions (generating schedules and production plans for lower levels).
- *Shop control*, performing real-time management of the tasks to be carried out and the resources available.
- *Cell control*, sequencing a batch of jobs through workstations and supervising materials handling.
- *Workstation control*, coordinating a small integrated group of shop-floor equipment during the execution of a task.
- *Equipment control*, supervising the actions of a particular piece of shop-floor equipment (e.g. a robot or a machine tool).

The distinguishing features of any hierarchical decomposition are as follows. In moving from higher levels to lower levels, the goals of the higher level are decomposed into commands sent to the lower levels. Feedback information from lower levels is passed back to the higher level, but at a decreasing rate (since information fed back at the higher level will be abstracted).

The development of intelligent manufacturing systems requires that all of these levels become more responsive to the current state of the system when they make their decisions. The particular focus of this chapter is on the *workstation* level of the AMRF hierarchy, which corresponds to what other authors define as the cell level. Argument as to what name to use is essentially sterile so it will suffice here to note that this chapter is concerned with *real-time* decision making at the *intermediate* levels of the factory hierarchy.

The activities addressed are those relating to the direct control (or coordination) of the systems concerned with the transformation of raw materials into products. The existence of low-level processing devices with known processing capabilities (the *equipment* level of the AMRF hierarchy) and higher-level systems (the *facility* and arguably the *shop* and *cell* levels of the AMRF hierarchy) acting as sources of driving information is assumed. Higher-level systems will feed production requirements to the real-time control system whose function is to coordinate the low-level computer-controlled machines carrying out the processing operations.

3.2   A Heterarchical Architecture

Hierarchical approaches are not universally accepted as the best way to tackle system design. Vamos (1983) and Duffie *et al.* (1986) argue that hierarchical approaches cannot deal with the combinatorial explosion of complexity presented by very-large-scale systems and that they are inflexible, difficult to expand, and unnecessarily complex. Consequently, there has been increasing interest over the last decade in what has been termed, by Hatvany (1984) and others, *heterarchic* manufacturing systems. The major difference between hierarchic and heterarchic organizations arises from the manner in which subsystems interact with one another. Instead of the binding control exercised by the higher levels over the lower levels of hierarchical systems, heterarchic subsystems exhibit what Enslow (1978) has termed *cooperative autonomy* in their interaction, whereby each entity is able to refuse a service request from another. Heterarchic systems are characterized as ones where there is no global information describing the state of the system, all information being distributed amongst the subsystems which base their decisions on local factors.

In comparing control software organized heterarchically against the conventional hierarchic approach Duffie and Piper (1986a,b) find the heterarchic approach to result in reduced complexity, reduced software cost, higher modularity, higher extensibility, and higher fault tolerance. The application domain for this work was in the control of manufacturing cells where distributed control ideas were applied to the coordination of the cell entities.

Further work on heterarchic approaches to the control of manufacturing systems, such as that reported in Parunak (1987a), Parunak *et al.* (1985), Shaw (1985, 1987), Maley and Solberg (1987a,b) and Upton (1988), has looked at the use of distributed control for the coordination of larger systems. Although promising results have been reported there is no conclusive evidence that the heterarchic approach will prove better than the hierarchic. It seems most likely that a hybrid approach, utilizing elements of both of the alternative approaches, will be most effective.

3.3   A Hybrid Approach

Despite the conflicting claims for the benefits of either the hierarchical or the heterarchical approach it seems likely that both have something to offer. Manufacturing systems must involve some form of hierarchy, particularly at the higher levels, due to the large range of response times required of different functional areas and the different levels of abstraction at which the system can be viewed from these levels. When considering the real-time decision-making required to operate the system at the level of the factory floor, however, it is less certain that hierarchical decomposition is applicable.

It is most likely that any real system will consist of a hybrid structure using hierarchical principles where they perform best and heterarchic ones where these are more appropriate (see Upton and Rogers (1991) for more discussion of this).

4   STATE-DRIVEN DECISION MAKING

Whether a hierarchical, heterarchical or hybrid architecture is utilized, the perspective taken of decision making and control in this chapter is a *state-driven* one. Any decision-making entity is considered to choose its *outputs* on the basis of the *inputs* to it and its current *state* (although the specification of this state may involve information about past performance or events). The requirements for a modelling technique to be used in this situation are only that it allows all three of these elements to be explicitly and dynamically represented and that it permits definition of the *decision process* whereby particular outputs are chosen given particular values of inputs and state.

It should be noted that there will be many options as to how to represent the input, state, and output of a system. Many of these will be equivalent in terms of their *information content* (i.e. the totality of information represented in any particular case), but they will probably differ in terms of their *efficiency* (the ease with which required information may be extracted from them), their *robustness* (the sensitivity of the representation to uncertain or unreliable information), and their ease of generation. Thus the ease of defining the system's decision process will probably vary across the representational options.

When choosing amongst such nominally equivalent representations it is thus important to consider the information that each makes *explicit* and that which is only available *implicitly* through further, potentially difficult, operations on the data. The choice of representation to use in any particular circumstances will thus depend on the characteristics of the control or decision-making task to be carried out.

5   MODELLING TECHNIQUES SURVEYED

A wide range of modelling approaches have been explored as applicable to the making of real-time control decisions in manufacturing systems. Some of these are surveyed and compared in the following subsections.

5.1   Finite-state Machines (FSM)

A finite-state machine (see Gill (1962) or Hopcroft and Ullman (1979) for a definition) can be used as a mathematical model of a system with discrete inputs and outputs. An FSM takes a string of input characters and outputs a string of output characters on a one-to-one basis, passing through a sequence of internal states in the process. At any point the output character chosen (and the next state changed to) depends on the current state of the machine (which can only take a finite set of values) and the next character input. Using this formalism to model a decision-making entity requires that the inputs and outputs of the system be considered as characters from the input and output alphabets of the machine and that all the possible states of the system can be enumerated and considered as internal states of the machine.

Finite-state machines form the basis for many of the control decisions in the AMRF which is modelled as a hierarchy of such machines. Albus *et al.* (1981) describe the reasons for adopting the formalism whilst Scott and Strouse (1984) report on its use at the workstation level. Within the AMRF both the input and output 'characters' comprise two sorts of information. The input comprises a command from the level above and some feedback from the level below, whilst the output consists of feedback to the level above and decomposed commands to the level below. The other factor involved in making a decision is the state of the level itself such that the choice of output depends both on the input and the current state whilst the action of making a decision also changes the state at this level.

Whilst the operation of such a decision maker is easily understood the formalism has limitations in the range of decisions it can make (see the following subsection on other formal language models). It can perform adequately in certain low-level situations with limited decision complexity but its performance degrades as the state space is increased.

An FSM's operation can be described in the form of a state stable where the next decision (in terms of output values and choice of next state) can be obtained by scanning the table for the current values of system state and input. In situations where the number of different states or the number of possible input values are large, the usefulness of this formalism deteriorates since every possible combination of state and input needs to be explicitly represented as a line in the table.

5.2   Formal Languages

Although FSMs have been discussed above in a separate subsection they should really be viewed as the least powerful of a range of techniques arising from the field of *formal languages* in computer science research (see Lewis and Papadimitriou (1981) or Hopcroft and Ullman (1979) for an introduction to formal languages). This field of research is concerned with the definition of languages and of machines capable of recognizing if a particular string belongs to a certain language. Four main classes of language (and corresponding machine) have been distinguished: regular grammar (finite-state automaton); context-free grammar (pushdown automaton); context-sensitive grammar (linear bounded automaton); unrestricted grammar (Turing machine).

The three additional classes offer increased modelling power in the sense that they are capable of expressing certain decisions that cannot be expressed using the FSM technique (in fact, the four techniques offer increasingly general modelling power such that the earlier named techniques are proper subsets of those named later). However, they are less easily understood than the FSM/state table formalism and would probably appear obscure to the typical engineering user who is unlikely to have had any exposure to them during formal training.

The use of such formal modelling techniques as the basis for the specification of control software offers potential benefits in solving the so-called *software crisis* currently besetting manufacturing systems whereby the development of software is the main bottleneck in the system design process. The use of formal techniques offers to decrease the lead time for software production and to improve the reliability of software so generated (see Naylor and Volz (1988) and Naylor and Maletz (1986) for more discussion of this).

There have been a number of interesting applications of formal languages in manufacturing reported in the literature. Upton and Barash (1988) have used context-free grammars in modelling the routing of components through a system. Context-free grammars have also been the basis of a system, reported by Joshi *et al.* (1991), for simplifying the software-generation process. The use of both context-free and context-sensitive grammar models of manufacturing control is discussed by Williams and Upton (1989) whilst Parunak (1987b) has shown that the scheduling problem for a general machine shop can be formulated as an unrestricted grammar and thereby shown to be amongst the most difficult type of problem.

5.3   Petri Nets

Petri nets (see Peterson (1981) for a description) can really be considered yet another dialect of formal language but they are given their own section here due to the enormous volume of literature on them and on their application in manufacturing. They were originally developed for the modelling of concurrent, asynchronous systems (systems involving a number of interacting, concurrently operating, semi-autonomous subsystems).

In the pure form they have a number of properties which make it possible to validate the logic they embody (e.g. to prove that the system design is deadlock free or that a particular desired state can be reached from the initial state). However, in their pure form they have proven to be unwieldy in modelling manufacturing systems of realistic complexity due to the number of possible states that such systems might occupy. An extension to the pure form known as 'coloured' Petri nets (see Alla *et al.* (1985) and Viswanadham and Narahari (1987) for representative papers in this area) allows this problem to be resolved but only at the cost of losing some of the desirable properties noted earlier.

One further feature of Petri nets in their purest form is that they do not model the flow of time explicitly. An extension to allow time modelling results in 'timed Petri nets' which have also seen much application (e.g. Ravichandran and Chakravarty (1986)) in the manufacturing area.

There have been a range of attempts to apply Petri net variants to manufacturing control problems at a range of levels from high-level planning (e.g. Grislain and Pun (1979) through cell coordination (e.g. Barad and Sipper (1988) and systems simulation (see Bruno and Morisio (1987) and Duggan and Browne (1988) to the definition of a programming language for programmable logic controllers (e.g. Murata *et al.* (1986) or for manufacturing cell controllers (see Crockett *et al.* (1987)).

5.4   Object-oriented Modelling

Object-oriented programming (see Cox (1986) for a thorough introduction) is another computer science development that has found widespread acceptance in the manufacturing research community. It has been heralded as a system building tool because of the modularity inherent in its approach.

A full description of this technique is given in a later section but to summarize its main features here, it is highly modular and allows computation to proceed by message passing between a number of separate entities, thus closely paralleling the real manufacturing world. A number of authors have explored the use of object-oriented programming in factory modelling including Reddy and Fox (1982), Stelzner *et al.* (1987), Bu-Hulaiga and Chakravarty (1988), King *et al.* (1988), Rogers and Williams (1988), Jarvinen and Konrad (1988), Adiga (1989), Adiga and Gadre (1990) and Glassey and Adiga (1990).

5.5   Knowledge-based Approaches

The use of artificial intelligence (AI) techniques for representing knowledge and decision making in manufacturing is a much researched area (see Kusiak (1988) and Pham (1988) for representative research). Work in this area is based on the use of such knowledge representation techniques as production rules, predicate logic, semantic networks and frames (see Barr and Feigenbaum (1983) for an introduction to each of these, Torsun (1984) for a brief comparison of them, and Brachmann and Levesque (1985) for a more detailed look at their history).

In particular the application of knowledge-based systems to various aspects of the scheduling problem has been widespread, for example Fox (1983), Yu and Wysk (1988), Ben-Arieh and Moodie (1987), Rogers *et al.* (1988), and Smith (1988).

Perhaps the most promising approach is that of developing hybrid modelling tools incorporating the best features of a number of the representational techniques noted above. An example of this is the work of Ruiz-Mier and Talavage (1987) who describe a formalism for intelligent manufacturing control derived from the diverse techniques of logic programming, object-oriented programming and functional programming.

The development of hybrid AI toolkits incorporating a range of knowledge representation paradigms has made work in this area easier to carry out. One particular problem such toolkits have been applied to is in the area of manufacturing simulation, resulting in what have been termed 'knowledge-based simulation' systems, for example Klahr and Faught (1980), Shannon (1988), Ben Arieh (1988), Acaccia *et al.* (1986), Sathi *et al.* (1987), Stelzner *et al.* (1987), Floss *et al.* (1986), Rogers and Williams (1988), and Widman *et al.* (1989).

5.6   Heterarchies or Distributed Agents

This approach is based on the idea that some systems are best controlled in a distributed fashion, where they may exhibit cooperative autonomy, rather than as rigid hierarchies. There has been some work on applying distributed control ideas to decision making in manufacturing systems based on the view of a system as a set of communication entities achieving coordination by negotiation. Much of this work is based on the concept of the contract net (as described in Smith (1980)) and its use to coordinate the negotiation process amongst a set of decision makers (see Davis and Smith (1983)).

These ideas have been applied in a number of domains such as small cells (e.g. Duffie and Piper, (1986a,b); (Duffie *et al.*, 1986, 1988)) and in the orchestration of larger systems (e.g. Maley and Solberg, (1987a,b); Maley, (1987)). Additionally, there has been some work by Parunak *et al.* (1985) attempting to apply Agha's (1986) concept of 'actors' to manufacturing control problems and by Ruiz-Mier and Talavage (1987) on the definition of a hybrid formalism for the specification of general manufacturing decision-making entities.

5.7   Connectionist Models

These models arise from research into possible mechanisms of control in the human brain where a very large number of relatively simple decision-making elements are connected together in the form of a neural net. It has been claimed by Smith *et al.* (1988) that neural networks are appropriate for the making of scheduling decisions because their features of being distributed, massively parallel, and continuously responsive mimic the requirements for real-time decision making in manufacturing. Although the applicability of distributed approaches to scheduling has been demonstrated, particularly in situations with significant volatility, there are doubts as to whether the scheduling task should be tackled in such a massively parallel way when significant hierarchical structuring of it is present.

Other applications rely on the ability of neural networks to learn how to solve a large class of problems by being exposed to a series of trial cases with answers given. An example of this type of work is reported by Chryssolouris *et al.* (1990) where a neural network is trained to select design parameters for a manufacturing system.

6   OBJECT-ORIENTED MODELLING

Stelzner *et al.* (1987) describe object-oriented programming (O-OP) as 'The process of building a computational model of a system consisting of discrete objects that interact with one another by passing messages . . .'. Programs written in object-oriented languages involve the definition of objects which store both the data and procedures required for the processing to be performed, computation proceeding via the passing of messages between such objects. A message can be viewed as a request for an information processing service to be carried out, manipulating the object data to produce a certain result. Such a request need not be concerned with the details of how the data is represented within the object providing the service nor with the details of how the manipulation is carried out.

The essence of O-OP is provided by the concepts of *encapsulation* and *inheritance* encourinreging, respectively, software modularity and the reuse of existing code modules in other applications. *Encapsulation*, the grouping (and possibly *hiding*) of data together with the procedures that manipulate it, encourages modularity by isolating the execution of any service to be carried out from the object asking for the service. Messages ask for services by name, and thus the sender need not be made aware of any changes made in the way the service is achieved. *Inheritance* promotes reuse of modules of code by allowing objects to be specified as specializations of other objects, such that only the way a new object differs from existing ones needs to be described. Thus new object descriptions may *inherit* the feature of previously defined objects, the final arrangement of objects forming a taxonomic hierarchy.

A typical *schema* (or object class definition) will consist of a set of *slots* storing attributes of the object concerned, and a set of *methods* defining how the object behaves (modifying its slots, or sending further messages) on receipt of messages. When an *instance* of this object class is created it will have its own values of all the defined slots (distinct from any other instance of the same class) but will respond to incoming messages in exactly the same way as other instances. The contents of each instance's slots will usually be *private*, that is inaccessible to, or hidden from, other objects except via methods.

The concept of O-OP is not a new one since its roots can be traced back to at least the early 1960s to the SIMULA language (see Birtwistle *et al.*, 1973, for details). There are a range of object-oriented languages (or object-oriented versions of common standard languages) now available, the most salient ones being C++ (see Stroustrup (1986) or Wiener and Pinson (1988)) and Smalltalk (see Pinson and Wiener (1988)). There have been many articles reporting systems developed using object-oriented languages across many application domains (see Pinson and Wiener (1990)) including manufacturing.

The use of O-OP offers benefits for software development in general in terms of enhanced modularity and flexibility and reduced requirements for new code. The inherent modularity of O-OP in particular makes object-oriented code easier to change (and thus easier to develop as the development process usually involves many changes in the specification of the software). O-OP also allows developers of code to work in terms of whatever level of abstraction they wish, defining object classes to represent whichever entities they discern in a particular situation (see Winblad *et al.* (1990)).

In terms of the modelling of real-world systems (and manufacturing systems in particular) O-OP languages provide the capability to represent the system *naturally*, developing software objects to correspond with the elements of the real system. Further, the intercommunication between such objects (whose similarities are recorded in an inheritance hierarchy, making them easier to define) can be partially represented by the *methods* defined to belong to the object classes. With such a scheme, a new entity can be added to a simulation model by defining an appropriate object class such that the attributes of the entity are stored as slots of the object whilst the entity's behaviour is represented as the methods (triggered by messages received) which modify this data. This approach has been explored extensively in object-oriented manufacturing simulation systems (e.g. see Reddy and Fox, (1982); Stelzner *et al.*, (1987); Rogers *et al.* (1990b); Glassey and Adiga, (1990)).

7   OBJECT-ORIENTED MODELS OF MANUFACTURING SYSTEMS

As was described above, an object in an object-oriented computer program consists of some private data and a collection of procedures that can access that data. Certain procedures are publicly accessible to other objects through the sending of a message to ask for a given service to be carried out. A manufacturing system corresponds to this in the sense that it consists of a collection of entities which communicate to ask tasks of each other in order to achieve the overall objective of the complete system.

There is a fundamental difference between the two, however, in that manufacturing systems are inherently parallel whilst a standard object-oriented computer program is inherently sequential. Within manufacturing systems there are a number of decision makers acting at the same time and possibly sending simultaneous (or nearly so) messages. The sender of a message is likely to continue carrying out its function after a message transmission, although it may need to wait for some sort of reply before continuing for more than a certain amount of time. In a conventional object-oriented computer program, however, the sender of a message waits for control to be passed back to it on completion of processing at the receiver before continuing (i.e. this involves synchronous messaging).

The principles of O-OP can be used, in a modified form, if this difference is explicitly acknowledged through the definition of a *manufacturing entity* as the fundamental building block from which full system models can be constructed. The material presented below, explaining the concept of the manufacturing entity, has similarities to arguments by Duffie *et al.* (1988) and to Zeigler's (1990) concept of the 'system entity structure'.

7.1   The Manufacturing Entity

The conception of the manufacturing entity stems from viewing a system as being state driven. All the decision-making entities comprising a manufacturing system are viewed as state-driven decision makers with the following characteristics.

- Each entity operates independently of and communicates asynchronously with the other entities comprising the system.
- Incoming communications or messages can be received directly and acted on according to the entity's internal decision-making logic.
- Actions can result in the generation of outgoing communications or messages which may be either broadcast or sent directly to other specific entities of which the entity is aware.
- Decisions are made solely on the basis of system state as represented in each entity's model of the system. This state incorporates the real-world state (as perceived via sensors), the

entity's internal state (as determined by some internal model), and the content of any incoming messages. Each entity's state model might be maintained either by the entity itself, as it communicates with the rest of the system, or independently of the entity with all relevant state changes being broadcast to it.

With this view, each decision maker is partially decoupled from the data on which it makes its decisions. Decision makers can thus operate in parallel whenever the system state is such that a decision needs to be made. Decision makers viewing the system at different levels can be accommodated, some operating according to fast control loops making repetitive decisions quickly whilst others base their decisions on aggregated data about the historical performance of the system. No restrictions are imposed on the structure that the actual decision-making element must take or on how it is described. Thus any of the formalisms introduced above might be used to specify the operation of a particular entity.

7.2 The Manufacturing Entity as a Low-level Device

This conceptual model can be used to describe the lowest-level controllers present in the manufacturing system, such as a computer numerical controller (CNC) or a programmable logic controller (PLC), in the following manner.

(1) The device runs a computer program controlling its actions which makes decisions on the basis of a model of the state of the device. This model includes elements describing the device internal state, the signals received from its sensors interfacing with the physical world, and communications received from other devices.

(2) Outgoing communications allow the reporting of a subset (or aggregation) of the device's local state, as modified by its local actions, to other devices to be used as inputs to their decision-making processes. Such reporting might be carried out at regular time periods or at the time of certain changes of state.

(3) The device can be considered as an autonomous entity, more or less so according to the amount of influence exerted over it by incoming messages. The more decisions are made without recourse to a higher authority, and the wider the range of outputs that the device can choose from in responding to the disturbances it is faced with, the greater is its autonomy.

It might be desirable to distinguish two subelements of a device's computer program, one being concerned with its actions in the real world and the other with the execution of these actions in response to communications from external devices (e.g. Duffie *et al.* (1988) distinguish *controller* and *communicator* elements for all entities involved in a heterarchical manufacturing cell). This latter element can be viewed as an 'executive' program allowing an external controller to start or stop the actions of the device or to access specific memory locations in the device's controlling computer (see Fussell *et al.* (1984) for a description of the communications requirements of a device controller for use in an automated manufacturing system). Such a distinction is not strictly necessary, however, as these executive actions can equally well be viewed as part of the normal state-driven operation of the device as supported by the manufacturing entity model.

7.3   The Manufacturing Entity as a Cell (Hierarchical/Heterarchical) Control Module

The main concern of this chapter is the development of real-time supervisory control software for manufacturing systems (i.e. the software to *coordinate* the operation of the low-level devices rather than to control their individual actions in detail). Such control can be effected by the definition of a set of control modules each taking the form of a manufacturing entity. The main point to note at this stage is that the decisions made by such an entity are made on the basis of a subset of the state of the overall system and are implemented by the passing of a message to other entities in the system.

The operation of a typical control module is described as follows. The module contains decision logic that continually monitors the state of the system, as reflected in a model of this (or more correctly in a model of a subset of the system state). When changes to this model so warrant, the control module will make a decision and generate a message to another entity to implement that decision. Alternatively, the control module might react to directly received messages if a quick response to a change in system state is required. It is possible to decouple the control module completely from the other entities by isolating it so that it cannot receive messages directly from other entities (i.e. no externally initiated messages are allowed). In this case, all decisions are made solely on the basis of the system state model which is maintained by entities broadcasting relevant state changes. Such a decoupling promotes the modular specification of the decision logic for the system since each independent decision can be distinguished and specified separately.

7.4   The Organization of Entities Comprising a System

The collection of entities defined as constituting the overall system may be arranged hierarchically or heterarchically, depending upon the decision-making logic each includes. This scheme can be compared with the architecture of the Advanced Manufacturing Research Facility (AMRF) introduced above. The manufacturing-entity-based architecture proposed here has three main advantages over the AMRF architecture, all involving a reduction in constraints imposed on the system. First, no hierarchical structure is imposed upon the system, so orders or commands can be contained in any incoming message, not strictly those which come from higher-level entities, allowing more cooperative decision making to be represented. Second, the state feedback allowed is more general since the maintenance of the overall state model is partially transparent to each decision-making entity; it need not come strictly from a lower-level entity and may not necessarily be a direct communication. Third, more complex decision making is allowed since the internal decision logic of a control module need not conform to the structure of a simple finite-state automaton.

With an architecture based on the concept of the manufacturing entity, new decision elements can be added as required if increased functionality is desired. Such decision elements need only see that portion of the overall state model (which can be considered as equivalent to a dynamic, perhaps distributed, database) that they require to see. The inclusion of the ability to react immediately to a received message permits certain states to be reacted to more quickly but directly couples the intercommunicating devices, rather than allowing a decision to be triggered solely by the state of the database.

8   UNRESOLVED PROBLEMS IN OBJECT-ORIENTED MANUFACTURING MODELLING

In using object-oriented modelling in manufacturing applications it is appealing to think of a manufacturing system model as involving communicating software objects, but care must be taken with the terminology! Although the entities discernible (or defined to be discernible) in an application can be thought of as objects in one sense of the word, they are *not* objects in the standard O-OP sense. O-OP and processing must be distinguished from truly distributed processing. An O-OP implementation of a program is not a distributed solution since there is a single thread of control running through the program; O-OP does break up the overall decision into semi-independent components but these do not act in parallel. As Cox (1986) is careful to point out in discussing the problems of developing distributed processing systems (of which manufacturing systems are an important type with the added complexity of carrying out actions in the physical world): 'Is object-oriented programming the answer? Absolutely not. It merely helps develop large, complicated, but basically conventional systems.'

A number of difficulties arise in attempting to model an inherently parallel system using an inherently sequential modelling paradism. The subsections below note four major problems which arise in the use of O-OP to describe general manufacturing systems together with possible solutions to some of these. The problems addressed are: modelling a system involving parallel decision makers on a single processor; allowing for a range of potential distributions of these decision makers amongst available computers in the real implementation; handling the distributed database representing the system state; allowing for the communication requirements of the real system at the modelling stage.

8.1   Modelling Parallel Decision Makers

Modelling a system consisting of parallel decision makers on a single, sequential processor is a general problem relevant to any simulation software. This might be resolved by specifying all the decision modules of interest in the model as state-driven entities and allowing each one to carry out as much processing as it requires whenever the system state is such that a decision must be made. For this approach to be valid, no single control module must be allowed to monopolize the resources of the single processor on which the collection of modules is implemented. This requires that some form of multi-tasking is provided to control the sharing of the single processing resource (whether this is built into or on top of the operating system).

8.2   Supporting a Distributed Implementation

The possible desire to distribute the independent decision makers present in a particular system across a number of physical processing devices presents unresolved problems, in the general case. Since conventional O-OP languages are inherently sequential, one possible solution here involves a careful integration of O-OP languages with discrete-event simulation techniques (as proposed in Ruiz-Micr and Talavage (1987) and Zeigler (1990)) to ensure that inherently parallel manufacturing systems are modelled in a valid manner. The enhanced O-OP models that result will perform three functions. The first is to represent the static and dynamic information describing the state of the system as an object-oriented database. The second is to define the simulated behaviour of the low-level devices comprising a system in a modular fashion. The third is to mirror a subset of the actual communications of the real system (though see below). This is a very active research area currently.

8.3   Handling a Distributed System Model

In the general case, a control system would involve a number of parallel decision makers distributed across a number of processing devices. In this situation problems of distributed databases arise (e.g. concerning consistency of duplicated information). This is another research area, related to the preceding subsection, where there is much activity at the present time.

8.4   Communications Modelling

The detailed modelling of much of the communications in a real system is a complex problem. There has been much activity in the past decade concerned with the definition of communications standards for manufacturing and other systems. Developers of control logic should be able to treat communications as a service to which they have access, without needing to consider the internal details of how the service is provided. This problem is not yet resolved, although ideas such as those of Duffie *et al.* (1988) referred to above, where each entity has distinct controller and communicator elements, go some way towards providing the necessary decoupling between the control and communication activities.

9   CONCLUSIONS

This chapter has focused on techniques for describing the real-time, state-driven decision-making logic that is needed to control manufacturing cells. Such techniques will be required independently of the architecture of a manufacturing system if it is to achieve a sufficient level of intelligence and responsiveness. A number of candidate techniques which have received the attention of researchers have been briefly surveyed and the object-oriented modelling paradigm has been highlighted as a particularly promising one. This technique offers the potential to support responsiveness not only in the sense of state-driven decision making but also in the sense of rapid development of control software. However, there are a number of unresolved issues that must be addressed before this potential can be reached.

REFERENCES

Acaccia G. M., Michelini R. C., Molfino R. M. and Piaggio P. A. (1986). X-SIFIP: a knowledge-based special purpose simulator for the development of flexible manufacturing cells. *Proc. IEEE Int. Conf. on Robotics and Automation*, pp. 645–63.

Adiga S. (1989). Software modelling of manufacturing systems: a case for an object-oriented programming approach. *Ann. Oper. Res.* 17, pp. 363–78.

Adiga S. and Gadre M. (1990). Object-oriented software modeling of a flexible manufacturing system. *J. Rob. Syst.*, to appear.

Agha G. A. (1986). *Actors: A Model of Concurrent Computation in Distributed Systems*. Cambridge, MA: MIT Press.

Albus J. S., Barbera A. J. and Nagel R. N. (1981). Theory and practice of hierarchical control. *23rd IEEE Computer Society Int. Conf., September, 1981*.

Albus J. S., Barbera A. J., Fitzgerald M. L., Kent R., McLean C., McCaih H., Bloom H., Haynes L., Furlani C., Barkmeyer E., Mitchell M., Scott H.,

Bloomquist D. and Kilmer R. *et al.* (1984). A control system for an automated manufacturing research facility. *Proc. Robots 8 Conf. and Exposition*, Detroit, MI, June, 1984.

Alla H., Ladet P., Martinez J. and Silva-Suarez M. (1985). Modelling and validation of complex systems by coloured Petri nets application to a flexible manufacturing system. In *Lecture Notes in Computer Science: Advances in Petri Nets 1984*. (Goos G. and Hartmanis J., eds) vol. 188, pp. 15–31.

Barad M. and Sipper D. (1988). Flexibility in manufacturing systems: definitions and Petri net modelling. *Int. J. Prod. Res.*, **26**, pp. 237–48.

Barr A. and Feigenbaum E. A. (eds) (1983). *The Handbook of Artificial Intelligence*, vol. I, Pitman.

Ben-Arieh D. (1988). A knowledge-based simulation and control system. In *Artificial Intelligence: implications for computer integrated manufacturing* (Kusiak A., ed.). IFS Publications and Springer, pp. 461–72.

Ben-Arieh D. and Moodie C. L. (1987). Knowledge based routing and sequencing for discrete part production. *J. Manuf. Syst.*, **6**, 287–97.

Birtwistle G. M., Dahl O.-J., Myhrhaug B. and Nygaard K. (1973). *SIMULA Begin*. Studentlitteratur.

Bourne D. A. and Fox M. S. (1984). Autonomous manufacturing: automating the job-shop. *IEEE Comput.*, **19**, 76–86, September.

Brachman R. J. and Levesque H. J. (eds) (1985). *Readings in Knowledge Representation*. Morgan Kaufmann.

Bruno G. and Morisio M. (1987). Petri-net based simulation of manufacturing cells. *Proc. IEEE Conf. on Robotics and Automation*, pp. 1174–9.

Bu-Hulaiga M. I. and Chakravarty A. K. (1988). An object-oriented knowledge representation for hierarchical real-time control of flexible manufacturing. *Int. J. Prod. Res.*, **26**, 777–93.

Chryssolouris G., Lee M., Pierce J. and Domroese M. (1990). Use of neural networks for the design of manufacturing systems. *Manuf. Rev.*, **3**, no. 3, pp. 187–194.

Cox B. J. (1986). *Object-oriented Programming: An Evolutionary Approach*. Addison-Wesley.

Crockett D., Desrochers A., DiCesare F. and Ward T. (1987). Implementation of a Petri net controller for a machining workstation. *Proc. IEEE Conf. on Robotics and Automation*, pp. 1861–7.

Davis R. and Smith R. G. (1983). Negotiation as a metaphor for distributed problem solving. *Artif. Intell.*, **20**, 63–109.

Duffie N. A. and Piper R. S. (1986a). Non-hierarchical control of a flexible manufacturing cell. *Proc. Int. Conf. on Intelligent Manufacturing Systems*, Budapest, Hungary, 16–19 June 1986.

Duffie N. A. and Piper R. S. (1986b). Nonhierarchical control of manufacturing systems. *J. Manuf. Syst.*, **5**, 137–9.

Duffie N. A., Piper R. S., Humphrey B. J. and Hartwick J. P. Jr (1986). Hierarchical and non-hierarchical manufacturing cell control with dynamic part-oriented scheduling. *Proc. NAMRC-XIV*, Minneapolis, MN, 28–30 May 1986.

Duffie N. A., Chitturi R. and Mou Jong-I. (1988). Fault-tolerant heterarchical control of heterogenous manufacturing system entities. *J. Manuf. Syst.*, 7, 315–28.

Duggan J. and Browne J. (1988). ESPNET: expert-system-based simulator of Petri nets. *IIE Proc.*, **135**, Pt D, no 4, pp. 239–247.

Enslow P. H. Jr (1978). What is a 'distributed' data processing system? *Computer*, pp. 13–21, January.

Floss P., Ruiz-Mier S. and Talavage J. (1986). *The SIMYON Manual*. Engineering Research Center for Intelligent Manufacturing Systems, Purdue University, Technical Report TR-ERC 86-19.

Fox M. S. (1983). *Constraint-Directed Search: A Case Study of Job-Shop Scheduling*. Carnegie-Mellon University, Robotics Institute, Technical Report, CMU-RI-TR-83-22.

Fussell P. S., Wright P. K. and Bourne D. (1984). A design of a controller as a component of a robotic manufacturing system. *J. Manuf. Syst.*, **3**, 1–11.

Gershwin S. B., Hildebrant R. R., Suri R. and Mitter S. K. (1986). A control perspective on recent trends in manufacturing systems. *IEEE Control Syst. Mag.*, April, pp. 3–15.

Gill A. (1962). *Introduction to the Theory of Finite-state Machines*. McGraw-Hill.

Glassey C. R. and Adiga S. (1990). Berkeley Library of Objects for Control and Simulation of Manufacturing (BLOCS/M). In *Applications of Object-Oriented Programming*, (Pinson L. J. and Wiener R. S., eds), Addison-Wesley.

Grislain J. A. and Pun L. (1979). Graphical methods for production control. *Int. J. Prod. Res.*, **17**, pp. 643–59.

Hatvany J. (1984). Intelligence and cooperation in heterarchic manufacturing systems. *Proc. 16th CIRP Int. Semin. on Manufacturing Systems*, Tokyo, 1984.

Hopcroft J. E. and Ullman J. D. (1979). *Introduction to Automata Theory, Languages, and Computation*. Addison-Wesley.

Jarvinen O. and Konrad H. (1988). Object-oriented modelling of production activity control systems. *Proc. 4th CIM Europe Conf., 18–20 May 1988*. IFS Publications and Springer, pp.353–64.

Joshi S. B., Mettala G. and Wysk R. A. (1991). CIMGEN – a computer aided software engineering tool for development of CIM software. *IIE Trans.*, to appear.

King C. U., Adams S. S. and Fisher E. L. (1988). Representation of manufacturing entities. In *Intelligent Manufacturing: Proc. First Int. Conf. on Expert Systems and the Leading Edge in Production Planning and Control* (Oliff M. D., ed.). Benjamin/Cummings, pp. 77–91.

Klahr P. and Faught W. S. (1980). Knowledge-based simulation. *Proc. First Annu. Conf. of the American Association for Artificial Intelligence*, pp. 181–3.

Kusiak A. (ed.) (1988). *Artificial Intelligence: implications for computer integrated manufacturing*. IFS (Publications) and Springer.

Lewis H. R. and Papadimitriou C. H. (1981). *Elements of the Theory of Computation*. Prentice-Hall.

Maley J. G. (1987). Part flow orchestration in distributed manufacturing processing. *PhD Thesis*, Purdue University.

Maley J. G. and Solberg J. J. (1987a). Managing the flow of intelligent parts. *Proc. Int. Conf. on the Manufacturing Science and Technology of the Future*, MIT, Cambridge, 3–7 June 1987.

Maley J. G. and Solberg J. J. (1987b). Part flow orchestration in CIM. *Proc. 9th Int. Conf. on Production Research*, Cincinnati, Ohio, 17–20 August 1987.

McLean C. R. (1985). An architecture for intelligent manufacturing control. *Proc. ASME Internal Conf. on Computers in Mechanical Engineering*, August, 1985.

Murata Tomohiro, Komoda Norihisa, Matsumoto Kuniaki and Haruna Koichi (1986). A Petri net-based controller for flexible and maintainable sequence control and its applications in factory automation. *IEEE Trans. Ind. Electron.*, **IE-33**, no. 1, pp. 1–8, February.

Naylor A. and Maletz M. C. (1986). The manufacturing game: a formal approach to manufacturing software. *IEEE Trans. Syst., Man and Cybern.*, **SMC-16**, 321–34.

Naylor A. W. and Volz R. A. (1988). Integration and flexibility of software for integrated manufacturing systems. In *Design and Analysis of Integrated Manufacturing Systems* (Dale Compton, W. Dale, ed.). Washington DC: National Academy Press.

Parunak H. V. D. (1987a). Manufacturing experience with the contract net. In *Distributed Artificial Intelligence*, vol. 1 (Huhns, M., ed.). London: Pitman.

Parunak H. V. D. (1987b). Why scheduling is hard (and how to do it anyway). *Proc. of the 1987 Materials Handling Focus (Research Forum)*, Georgia Institute of Technology, September 1987.

Parunak H. V. D., Irish B. W., Kindrick J. and Lozo P. W. (1985). Fractal actors for distributed manufacturing control. *Proc. 2nd IEEE Conf. on Artificial Intelligence Applications*, pp. 653–60.

Peterson J. L. (1981). *Petri Net Theory and the Modeling of Systems*. Englewood Cliffs, NJ: Prentice-Hall.

Pham D. T. (1988). *Expert Systems in Engineering*. IFS Publications and Springer.

Pinson L. J. and Wiener R. S. (1988). *An Introduction to Object-Oriented Programming and Smalltalk*. Addison-Wesley.

Pinson L. J. and Wiener R. S. (1990). *Applications of Object-Oriented Programming*. Addison-Wesley.

Ravichandran R. and Chakravarty A. K. (1986). Decision support in flexible manufacturing systems using timed Petri nets. *J. Manuf. Syst.*, **5**, no. 2, pp. 89–101.

Reddy Y. V. and Fox M. S. (1982). *KBS: An Artificial Intelligence Approach to Flexible Simulation*. Carnegie-Mellon University Robotics Institute Technical Report, CMU-RI-TR-82-1.

Rogers P. and Williams D. J. (1988). A knowledge-based system linking simulation to real-time control for manufacturing cells. *Proc. IEEE Int. Conf. on Robotics and Automation*, Philadelphia, 25–29 April 1988, pp. 1291–3.

Rogers P., Williams D. J., Wesley P. S. and Clare J. N. (1988). On-line scheduling of machining cells using knowledge-based simulation. *Proc. 4th Int. Conf. on Simulation in Manufacturing*, 2–3 November 1988, pp. 151–63.

Rogers P., Upton D. M. and Williams D. J. (1990a). Manufacturing, integration and computers: an engineering view of CIM. *Harvard Working Paper*, no. 91-023, October 1990.

Rogers P., Williams D. J. and Wesley P. S. (1990b). Object-oriented modelling for the design and scheduling of flexible machining cells incorporating a tool management system. *Proc. 1st Int. Conf. on Artificial Intelligence and Expert Systems in Manufacturing*, 20–21 March 1990.

Ruff K. (1985). Contemporary manufacturing systems integration. *Proc. National Science Foundation, University of Michigan Workshop on Systems Integration Tools in Manufacturing*, St Clair, Michigan, November 1985.

Ruiz-Mier S. and Talavage J. (1987). *A hybrid representation paradigm for intelligent manufacturing systems*. Engineering Research Center for Intelligent Manufacturing Systems, Purdue University, Technical Report TR-ERC 87-8.

Sathi N., Fox M. S., Baskaran V. and Bouer J. (1987). *An Artificial Intelligence Approach to the Simulation Life Cycle*. Carnegie Group Technical Brief.

Scott H. and Strouse K. (1984). Workstation control in a computer integrated manufacturing system. *Proc. Autofact* 6 Anaheim, CA, 4 October 1984.

Shannon R. E. (1988). Knowledge based simulation techniques for manufacturing. *Int. J. Prod. Res.*, **26**, 953–74.

Shaw M. J. (1985). Task bidding and distributed planning in flexible manufacturing. *Proc. 2nd Conf. on Artificial Intelligence Applications, IEEE, Miami Beach, FL*, pp. 184–9.

Shaw M. J. (1987). A distributed scheduling method for computer integrated manufacturing: the use of local area networks in cellular systems. *Int. J. Prod. Res.*, **25**, 1285–303, September.

Singh M. G. (1980). *Dynamic Hierarchical Control*, revised edition. North-Holland.

Smith A. W., Baghernejad M. and Stevenson I. A. (1988). Architectures for intelligent real-time control systems for manufacturing applications. *Proc. Int. Conf. on Factory 2000: Integrating Information and Material Flow*, Cambridge, UK, 31 August–2 September 1988, pp. 67–74.

Smith R. G. (1980). The contract net protocol: high-level communication and control in a distributed problem solver. *IEEE Trans. Comput.*, **C-29**, 104–13.

Smith S. F. (1988). A constraint based framework for reactive management of factory schedules. In *Intelligent Manufacturing: Proc. First Int. Conf. on Expert Systems and the Leading Edge in Production Planning and Control* (Oliff M. D., ed.). Benjamin/Cummings, pp. 113–30.

Stelzner M., Dynis J. and Cummins F. (1987). *The SimKit System: Knowledge-Based Simulation and Modeling Tools in KEE*, An Intellicorp Technical Article.

Stroustrup B. (1986). *The C++ Programming Language*. Addison-Wesley.

Torsun I. S. (1984). Knowledge representation – an overview. *Proc. Second Workshop on Architectures for Large Knowledge Bases*, Manchester University, 9–11 July 1984.

Upton D. M. (1988). The operation of large computer controlled manufacturing systems. *PhD Thesis*, Purdue University.

Upton D. M. and Barash M. M. (1988). A grammatical approach to routing flexibility in large manufacturing systems. *J. Manuf. Syst.*, 7, 209–21.

Upton D. M. and Rogers P. (1991). Hierarchies, heterarchies and hybrids: architectural issues in automated manufacturing systems. *TIMS/ORSA Spring Meet.*, Nashville, TN, May 1991.

Vamos T. (1983). Cooperative systems – an evolutionary perspective. *IEEE Control Syst. Mag.*, pp. 9–14, August.

Viswanadham N. and Narahari Y. (1987). Coloured Petri nets models for automated manufacturing systems. *Proc. IEEE Conf. Robotics and Automation*, pp. 1985–90.

Widman L. E., Loparo K. A. and Nielsen N. R. (eds) (1989). *Artificial Intelligence, Simulation and Modeling*. Wiley.

Wiener R. S. and Pinson L. J. (1988). *An Introduction to Object-Oriented Programming and C++*. Addison-Wesley.

Williams D. J. and Upton D. M. (1989). Syntactic models of manufacturing processing and control. *Int. J. CIM*, 2, 229–37.

Winblad A. L., Edwards S. D. and King D. R. (1990). *Object-Oriented Software*. Addison-Wesley.

Yu S.-Y. and Wysk R.A. (1988). Multi-pass expert system control system – a control/scheduling structure for flexible manufacturing cells. *J. Manuf. Syst.*, 7, 107–20.

Zeigler B. P. (1990). *Object-Oriented Simulation with Hierarchical, Modular Models: Intelligent Agents and Endomorphic Systems*. Academic Press.

Index